# Springer Proceedings in Complexity

Springer Proceedings in Complexity publishes proceedings from scholarly meetings on all topics relating to the interdisciplinary studies of complex systems science. Springer welcomes book ideas from authors. The series is indexed in Scopus.

Proposals must include the following:

- name, place and date of the scientific meeting
- a link to the committees (local organization, international advisors etc.)
- scientific description of the meeting
- list of invited/plenary speakers
- an estimate of the planned proceedings book parameters (number of pages/articles, requested number of bulk copies, submission deadline)

Submit your proposals to: christoph.baumann@springer.com

More information about this series at http://www.springer.com/series/11637

Dariusz Grech · Janusz Miśkiewicz
Editors

# Simplicity of Complexity in Economic and Social Systems

Proceedings of the 54th Winter School of Theoretical Physics, Lądek Zdrój, Poland, February 18–24th 2018

Springer

*Editors*
Dariusz Grech
Institute of Theoretical Physics
University of Wrocław
Wrocław, Poland

Janusz Miśkiewicz
Institute of Theoretical Physics
University of Wrocław
Wrocław, Poland

ISSN 2213-8684 ISSN 2213-8692 (electronic)
Springer Proceedings in Complexity
ISBN 978-3-030-56159-8 ISBN 978-3-030-56160-4 (eBook)
https://doi.org/10.1007/978-3-030-56160-4



This Springer imprint is published by the registered company Springer Nature Switzerland AG
The registered company address is: Gewerbestrasse 11, 6330 Cham, Switzerland

# Preface

Institute of Theoretical Physics, University of Wrocław supported by the Polish Physical Society Section "Physics in Economy and Social Sciences" (FENS) organized in February 2018 the 54th International Winter School of Theoretical Physics. The school entitled "Simplicity of Complexity in Economic and Social Systems" was mainly dedicated for graduated students, Ph.D. students and post-docs. Our goal was to make popularization of econophysics and sociophysics among world community of the younger generation of researchers.

International Winter Schools systematically organized by the Institute of Theoretical Physics have already had a very long tradition for they have been organizing yearly for 55 years now and they are well recognizable in the community of physicists. The details of all schools organized so far can be found at http://www.ift.uni.wroc.pl/conferences/list/type/Karpacz.

We do observe that the young generation is more and more open to interdisciplinary view at science. In this spirit, new interdisciplinary research programme was established at the end of twentieth century which links statistical physics (in particular, physics of complex systems) with financial analysis and sociology. In this way, the new areas of research appeared in science and computer science like econophysics (called also as financial physics) and sociophysics.

The main tools used in these areas are numerical simulation of agents' behaviour and the interpretation of numerical and analytical results of such simulation with the help of complexity tools. To do so the knowledge having its background in statistical physics and in physics of phase transitions is extremely helpful. This way physicists are trying to make contribution to better understanding of various phenomena in quantitative way in so fast-changing world around us. The number of advanced conferences in econophysics and sociophysics is thus still increasing—just to mention the biggest periodic ones like International Conference on Econophysics (see, e.g. http://ice2017.csp.escience.cn/dct/) or Econophysics Colloquium series (see, http://www.ec2017.org/, https://sites.google.com/view/econophysics-colloquium-2018 for two last meetings). However, there still exists a gap between these advanced conferences and strictly training events on econophysics and sociophysics organized for younger scientists where they may get their excellence in research skills and learn how ideas of physics can be successfully employed in other areas. Younger researchers can simply become fascinated during such events with a new interdisciplinary theme created by physicists, economists, sociologists and IT professionals. All this helps to understand better the complexity phenomena existing not only in physics but also in complex systems being seemingly far from traditional view at physics.

The event we have organized is exactly an advanced training meeting for younger participants—not a conference nor a symposium usually designed to publish recent (often unpublished) research. We feel that the School gave younger scientists and Ph.D. students an excellent preparation for participation in more advanced econophysics and sociophysics events like the forthcoming (in time this preface is written) Econophysics Colloquium 2019 organized in Singapore or similar events connected with application of complex systems analysis outside physics. The success of our School was guaranteed by prominent lecturers whose scientific achievements in econo- and sociophysics and teaching skills are very well confirmed and who agreed to come to Poland. They formed also the Scientific Committee of our training conference.

This book contains the chosen lectures from the meeting. We are proud to publish them in Proceedings of Complexity series by Springer and we do hope that this way our goal of reaching the conference lectures to the widest possible group of interested people would become true with benefit for all—researchers, students and Ph.D. students working on multidisciplinary approaches in science. We—the organizers—would also like to thank our sponsors—first of all the Ministry of Science and Higher Education for the financial support of DUN project under the contract no. 862/P-DUN/2018 and to the Rector of Wrocław University for the special assistance.

Wrocław, Poland

Editors of Proceedings
Dariusz Grech
Janusz Miśkiewicz

# Contents

# Contributors

**Marcel Ausloos** School of Business, University of Leicester, Leicester, UK; Department of Statistics and Econometrics, Bucharest University of Economic Studies, Bucharest, Sector 1, Romania;
GRAPES. rue de la Belle Jardiniere, Liege Angleur, Belgium

**Francesca Bartolacci** Department of Economics and Law, University of Macerata, Macerata, Italy

**Nicola G. Castellano** Department of Economics and Management, University of Pisa, Pisa, Italy

**Roy Cerqueti** Sapienza University of Rome, Department of Social and Economic Sciences, Rome, Italy;
London South Bank University, School of Business, London, UK

**Łukasz Gajewski** Faculty of Physics, Warsaw University of Technology, Warsaw, Poland

**Tomasz Gubiec** Faculty of Physics, University of Warsaw, Warsaw, Poland

**Janusz A. Hołyst** Faculty of Physics, Warsaw University of Technology, Warsaw, Poland;
ITMO University, Saint Petersburg, Russia

**Jarosław Klamut** Faculty of Physics, University of Warsaw, Warsaw, Poland

**Adam Krawiec** Institute of Economics, Finance and Management, Jagiellonian University, Kraków, Poland

**Ladislav Kristoufek** Institute of Information Theory and Automation, Czech Academy of Sciences, Prague, Czech Republic

**Ryszard Kutner** Faculty of Physics, University of Warsaw, Warsaw, Poland

**Paweł Oświęcimka** Complex Systems Theory Department, Institute of Nuclear Physics Polish Academy of Sciences, Krakow, Poland

**Robert Paluch** Faculty of Physics, Warsaw University of Technology, Warsaw, Poland

**Krzysztof Suchecki** Faculty of Physics, Warsaw University of Technology, Warsaw, Poland

**Marek Szydłowski** Astronomical Observatory, Jagiellonian University, Kraków, Poland;
Mark Kac Complex Systems Research Centre, Jagiellonian University, Kraków, Poland

**Bolesław Szymański** Social Cognitive Networks Academic Research Center, Rensselaer Polytechnic Institute, Troy, NY, USA;
Społeczna Akademia Nauk, Łódź, Poland

# Chapter 1
# Simple Approaches on How to Discover Promising Strategies for Efficient Enterprise Performance, at Time of Crisis in the Case of SMEs: Voronoi Clustering and Outlier Effects Perspective

**Marcel Ausloos, Francesca Bartolacci, Nicola G. Castellano, and Roy Cerqueti**

**Abstract** This paper analyses the connection between innovation activities of companies—implemented before a financial crisis—and their performance—measured after such a time of crisis. Pertinent data about companies listed in the STAR Market Segment of the Italian Stock Exchange is analysed. Innovation is measured through the level of investments in total tangible and intangible fixed assets in 2006–2007, while performance is captured through growth—expressed by variations of sales or of total assets—profitability—through ROI or ROS evolution—and

---

M. Ausloos (✉)
School of Business, University of Leicester, Brookfield, LE2, 1RQ Leicester, UK

Department of Statistics and Econometrics, Bucharest University of Economic Studies, Calea Dorobantilor 15-17, Bucharest 010552, Sector 1, Romania

GRAPES. rue de la Belle Jardiniere, 483/0021 B-4031 Liege Angleur, Belgium
e-mail: ma683@le.ac.uk; marcel.ausloos@ase.ro; marcel.ausloos@ulg.ac.be

F. Bartolacci
Department of Economics and Law, University of Macerata, Via Crescimbeni, 20, 62100 Macerata, Italy
e-mail: francesca.bartolacci@unimc.it

N. G. Castellano
Department of Economics and Management, University of Pisa, Via Ridolfi 10, 56124 Pisa, Italy
e-mail: nicola.castellano@unipi.it

R. Cerqueti
Sapienza University of Rome, Department of Social and Economic Sciences, Rome I-00185, Italy
e-mail: roy.cerqueti@uniroma1.it; cerquetr@lsbu.ac.uk

London South Bank University, School of Business, London SE1 0AA, UK


D. Grech and J. Miśkiewicz (eds.), *Simplicity of Complexity in Economic and Social Systems*, Springer Proceedings in Complexity,
https://doi.org/10.1007/978-3-030-56160-4_1

productivity—through asset turnover or sales/employee in the period 2008–2010. The variables of interest are analysed and compared through statistical techniques and by adopting a cluster analysis. In particular, a Voronoi tessellation is implemented in a varying centroids framework. In accord with a large part of the literature, we find that the behaviour of the performance of the companies is not univocal when they innovate. The statistical outliers are the best cases in order to suggest efficient strategies. In brief, it is found that a positive rate of investments is preferable.

## Introduction

This chapter is based on three recent papers:

- ([6]) F. Bartolacci, N.G. Castellano, and R. Cerqueti (2015). The impact of innovation on companies' performance: an entropy-based analysis of the STAR Market Segment of Italian Stock Exchange, Technology Analysis and Strategic Management 27, 102–123.
- ([3]) M. Ausloos, F. Bartolacci, N.G. Castellano, and R. Cerqueti (2018). Exploring how innovation strategies at time of crisis influence performance: a cluster analysis perspective, Technology Analysis and Strategic Management 30, 484–497.
- ([4]) M. Ausloos, R. Cerqueti, F. Bartolacci, and N. G Castellano (2018). SME investment best strategies. Outliers for assessing how to optimize performance, Physica A 509, 754–765.

The connection between innovation strategies (usually taken as the investments) of companies if implemented before a financial crisis and their performance measured after crisis time are interesting aspects of small and medium size enterprises (SME) economic life.

In fact, Latham and Braun [29], in *"Economic recessions, strategy, and performance: a synthesis"* claimed that

> Despite the episodic pervasiveness of recessions and their destructive impact on firms, a void exists in the management literature examining the intersection between recessions, strategy, and performance.

Therefore, it seems worthwhile to reflect on such connections considering practical cases. Thus, we have considered companies listed in the STAR Market Segment of the Italian Stock Exchange in recent times. SME innovation is here below measured through the level of investments in total tangible and intangible fixed assets in [2006–2007], while performance is captured through (i) growth—expressed by variations of sales (DS) and variations of total assets (DA); (ii) profitability—through returns on investments (ROI) and returns on sales (ROS) and (iii) productivity—through asset turnover (ATO) or sales per employee (S/E) in the period [2008–2010].

In the Milano STAR market, 71 companies of mid-size are listed, at the time of study: their capitalization value was between 40 million and 1 billion euros. Since their activity and innovation levels are different from "industrial companies", whence since their performance should be measured in a different way, we have removed banks and insurance institutes from our analysis. Thus, in the following, the segment is reduced to 62 SMEs.[1] For completeness, the 62 SMEs at the time of study are given in Table 1.1.

We discuss a formal method, based on Voronoi tessellation (Voronoi, 1908), yet we depart from the original formulation of Voronoi by introducing a concept of weighted Euclidean distances, hence leading to asymmetry (see formulas (2) and (3)). In our approach, we a priori define some reference points—so-called “centroids”, each centroid identifying a cluster whose elements are at a distance smaller than that of the other centroids.

For more information, let us mention that the use of Voronoi tessellation can be found in [32, 46, 48].

Such a cluster analysis is employed to investigate the determinants of innovation and innovation-performance focused on a single industry [45] or on different industries [10, 31, 38].

## Data

A few notations are to be introduced for easy readability of the following tables and figures:

- TIAXyy represents the level of total intangible assets (excluding goodwill) in year 20yy.
- TTAyy is the level of total tangible assets (excluding properties) in 20yy.
- DSyy stands for sales variations in year 20yy.
- DAyy is total assets variations in year 20yy.
- ROIyy means the return on investments in year 20yy.
- ROSyy means the return on sales in year 20yy.
- ATOyy represents asset turnover in year 20yy.
- S/Eyy stands for sales per employee in year 20yy.

- The lowest TTA value is called TTA1, while the highest TTA is TTA2.
- Their average is <TTA>2 = (1/2) (TTA1 + TTA2)

which, in fact, due to the time interval of interest, is equal to (TTA06 + TTA07). Similarly,

- <TIAX>2 is the average total intangible asset (excluding goodwill) over 2 years: [2006–2007]; “obviously”, <TIAX>2 = (1/2) (TIAX1 + TIAX2) = (1/2) (TIAX06 + TIAX07).

[1] The below displayed data can be obtained from the authors upon request as Excel tables.

**Table 1.1** The 62 STAR company names, at the time of study; alphabetical order and the "supersector" to which they belong; "supersector" abbreviations: Automobiles & Parts (A&P); Construction & Materials (C&M); Industrial Goods & Services (IG&S); Personal & Household Goods (P&HG); Food & Beverage (F&B)

| $i =$ | | Supersector | $i =$ | | Supersector |
|---|---|---|---|---|---|
| 1 | Acotel Group | Telecommunications | 32 | Exprivia | Technology |
| 2 | Aeffe | P&HG | 33 | Falck Renewables | Utilities |
| 3 | Amplifon | Health Care | 34 | Fidia | IG&S |
| 4 | Ansaldo Sts | IG&S | 35 | Fiera Milano | IG&S |
| 5 | Ascopiave | Utilities | 36 | Gefran | IG&S |
| 6 | Astaldi | C&M | 37 | I.M.A | IG&S |
| 7 | Biancamano | IG&S | 38 | Interpump Group | IG&S |
| 8 | Biesse | IG&S | 39 | Irce | IG&S |
| 9 | Bolzoni | A&P | 40 | Isagro | Chemicals |
| 10 | Brembo | A&P | 41 | It Way | Technology |
| 11 | Buongiorno * | Technology | 42 | La Doria | F&B |
| 12 | Cad It | Technology | 43 | Landi Renzo | A&P |
| 13 | Cairo Communic. | Media | 44 | Marr | Retail |
| 14 | Cembre | IG&S | 45 | Mondo Tv | Media |
| 15 | Cementir Holding | C&M | 46 | Nice | IG&S |
| 16 | Centrale Latte To | F&B | 47 | Panariagroup | C&M |
| 17 | Cobra | Automobiles and Parts | 48 | Poligraf. S. F | IG&S |
| 18 | Dada | Technology | 49 | Poltrona Frau | P&HG |
| 19 | Damiani | Retail | 50 | Prima Industrie | IG&S |
| 20 | D'Amico | IG&S | 51 | Rdb | C&M |
| 21 | Datalogic | IG&S | 52 | Reno De Medici | IG&S |
| 22 | Digital Bros | P&HG | 53 | Reply | Technology |
| 23 | Dmail Group | Media | 54 | Sabaf | IG&S |
| 24 | Dmt | Technology | 55 | Saes Getters | IG&S |
| 25 | Eems ** | Technology | 56 | Servizi Italia | IG&S |
| 26 | El.En | IG&S | 57 | Sogefi | A&P |
| 27 | Elica | IG&S | 58 | Ternienergia | Utilities |
| 28 | Emak | P&HG | 59 | Tesmec | IG&S |
| 29 | Engineering | Technology | 60 | Txt E-Solutions | Technology |
| 30 | Esprinet | Technology | 61 | Yoox | Retail *** |
| 31 | Eurotech | Technology | 62 | Zignago Vetro | IG&S |

*Since July 2012, Buongiorno is part of Docomo Digital

**Eems was moved away from Technology in STAR to MTA Market/Segment

***In March 2015, Yoox merged with Net-a-Porter

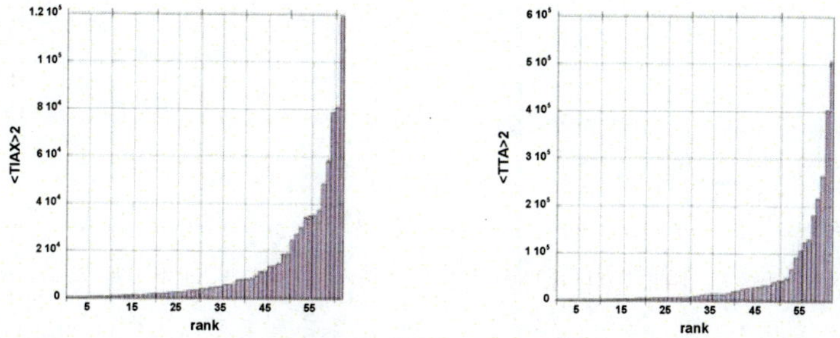


**Fig. 1.1** Left panel: <TIAX>2. Right panel: <TTA>2; thus each averaged over 2 years: [2006–2007]; both data are ranked in increasing order—for the 62 SMEs discussed in the text

**Table 1.2** Main statistical indicators of the innovation and performance variables; $Tot.Ass.$ = "Total Assets"; $Var.n$ = "variation"; *empl.* = "employee"; $Ret.on$ = returns on

| | Innovation | | Performance | | | | | |
|---|---|---|---|---|---|---|---|---|
| | | | Growth | | Profitability | | Efficiency | |
| | Intangible Assets (TIAX) | Tangible Assets (TTA) | Sales $Var.n$ (DS) | $Tot.Ass.$ $Var.n$ (DA) | $Ret.on$ Invest. (ROI) | $Ret.on$ Sales (ROS) | Asset turnover (ATO) | Sales/ *empl.* (S/E) |
| Mean ($\mu$) | 12,360.46 | 29,215.40 | 9% | 6% | 5% | 5% | 0,91 | 275.77 |
| Std.Dev.($\sigma$) | 18,695.11 | 45,379.80 | 16% | 14% | 5% | 7% | 0,34 | 231.20 |
| $\mu/\sigma$ | 0.66 | 0.64 | 0,46 | 0,57 | 0,85 | 0,75 | 2,68 | 1.19 |
| min. | 180 | 86.50 | −19% | −10% | −8% | −14% | 0,15 | 57.20 |
| Max | 80,816 | 217,237.50 | 59% | 53% | 21% | 24% | 2,04 | 1,100.76 |
| Q1 | 1,346.50 | 3,579.50 | −1% | −2% | 2% | 1% | 0,75 | 148.02 |
| Median | 3,584 | 10,329.00 | 3% | 4% | 4% | 5% | 0,86 | 188.04 |
| Q3 | 13,917 | 31,331.50 | 12% | 16% | 8% | 9% | 1,09 | 281.07 |
| Skewness | 2.28 | 2.57 | 1.48 | 1.32 | 0.44 | 0.27 | 0.78 | 2.25 |
| Kurtosis | 5.14 | 6.79 | 3.60 | 1.07 | 0.60 | 0.76 | 1.76 | 5.05 |

We provide figures in order to visualize the data range for <TIAX>2 and <TTA>2 shown in Fig. 1.1. In these figures, the SMEs are ranked in increasing order of the $y$-variable value. The range and statistical characteristics are outlined in Table 1.2. Other displays, e.g. when the SMEs are listed in alphabetical order, on the $x$-axis can be found in Fig. 1.2.

Next, let us display the performance variables averaged over 3 years, [2008–2010]:

- <DS>3 for the sales variations,
- <DA>3 for the total assets variations,
- <ROI>3 for ROI,
- <ROS>3 for ROS,
- <ATO>3 for the asset turnovers and

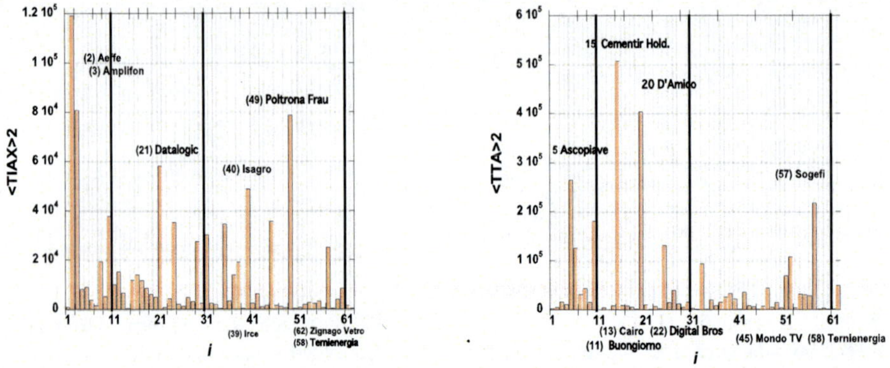


**Fig. 1.2** (colour online) Left panel: <TIAX>2. Right panel: <TTA>2; thus each averaged over 2 years: [2006–2007],—for the 62 SMEs, ranked in alphabetical order as in Table 1.1, particularly pointing to a few relevant SMEs of the STAR market so studied

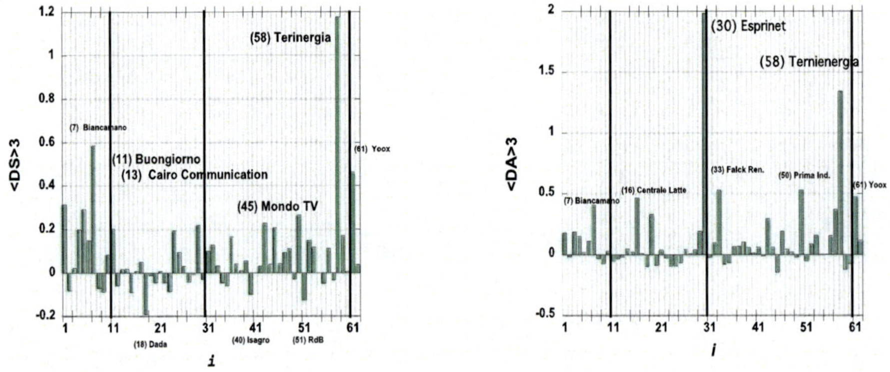


**Fig. 1.3** (colour online) Left panel: sales variations <DS>3. Right panel: total assets variations <DA>3; thus each averaged over 3 years: [2008–2010]—for the 62 SMEs, ranked in alphabetical order as in Table 1.1, particularly pointing to a few relevant SMEs of the STAR market so studied

- <S/E>3 for the sales per employee,

either when companies are listed in alphabetical order, as in Figs. 1.3, 1.4 and 1.5, or ranked in increasing order of the relevant variable, as in Figs. 1.6, 1.7 and 1.8.

Statistical characteristics for the distributions of the averaged innovation and performance indicators are found in Table 1.3.

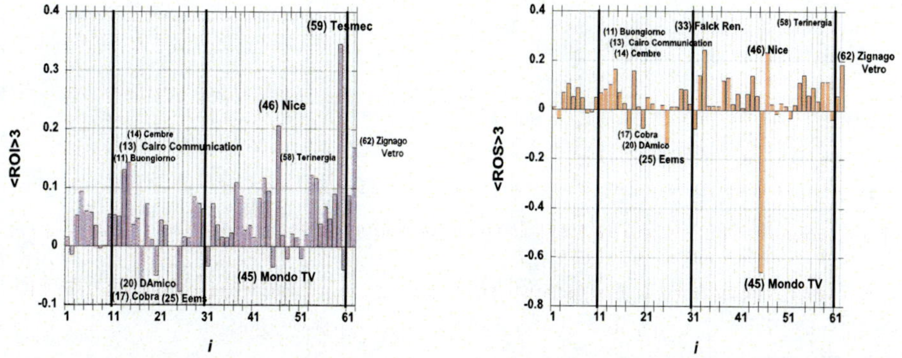


**Fig. 1.4** (colour online) Left panel: returns on investments <ROI>3. Right panel: returns on sales <ROS>3. Thus each averaged over 3 years: [2008–2010]—for the 62 SMEs, ranked in alphabetical order as in Table 1.1, particularly pointing to a few relevant SMEs of the STAR market so studied

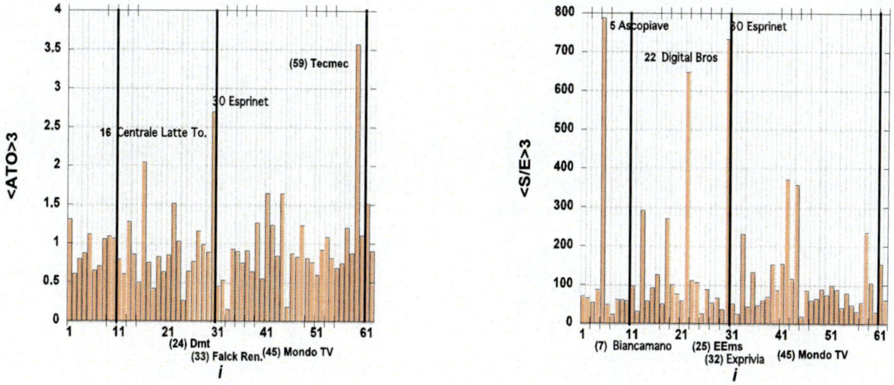


**Fig. 1.5** (colour online) Left panel: asset turnovers <ATO>3. Right panel: sales per employee <S/E>3. Thus each averaged over 3 years: [2008–2010]—for the 62 SMEs, ranked in alphabetical order as in Table 1.1, particularly pointing to a few relevant SMEs of the STAR market so studied

## Discussion

Many correlations can be searched for, besides those[2] between TTA06 and TTA07, or TIAX06 and TIAX07, shown in Fig. 1.9, one may consider those between the averaged variables, like

- <DS>3 versus <TTA>2
- <DA>3 versus <TTA>2
- <ROI>3 versus <TTA>2
- <ROS>3 versus <TTA>2

[2]Notice that the relationships are not exactly linear.

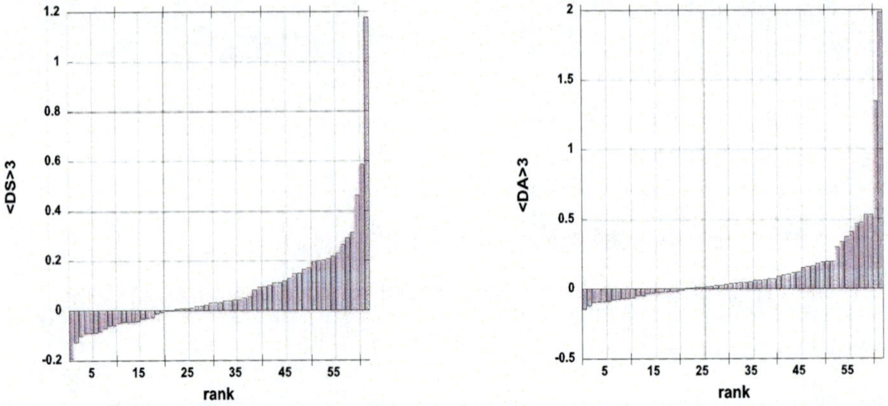


**Fig. 1.6** (colour online) Left panel: sales variations <DS>3. Right panel: total assets variations <DA>3; both ranked in increasing order—for the 62 SMEs, particularly pointing to a few relevant SMEs of the STAR market so studied

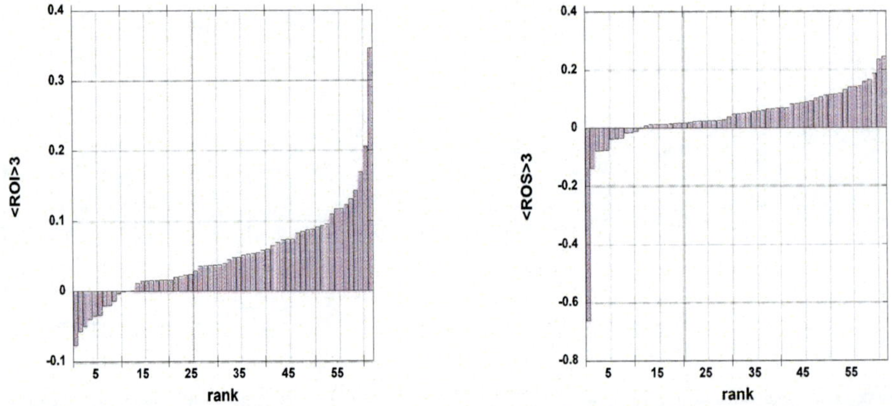


**Fig. 1.7** (colour online) Left panel: returns on investments <ROI>3. Right panel: returns on sales <ROS>3; both ranked in increasing order—for the 62 SMEs, particularly pointing to a few relevant SMEs of the STAR market so studied

which can be read in Figs. 1.8, 1.9, 1.10 and fig:11, in [4], whence are not reproduced here. Nevertheless, for completeness, we show

- <ATO>3 versus <TIAX>2
- <S/E>3 versus <TIAX>2
- <ATO>3 versus <TTA>2
- <S/E>3 versus <TTA>2

on Figs. 1.10 and 1.11.

It should be apparent that the data looks pretty scattered, suggesting a "more sophisticated" approach for reaching some conclusion. As an intermediary remark, observe that <ATO>3 and <S/E>3 are all positive; this is not the case for <DS>3,

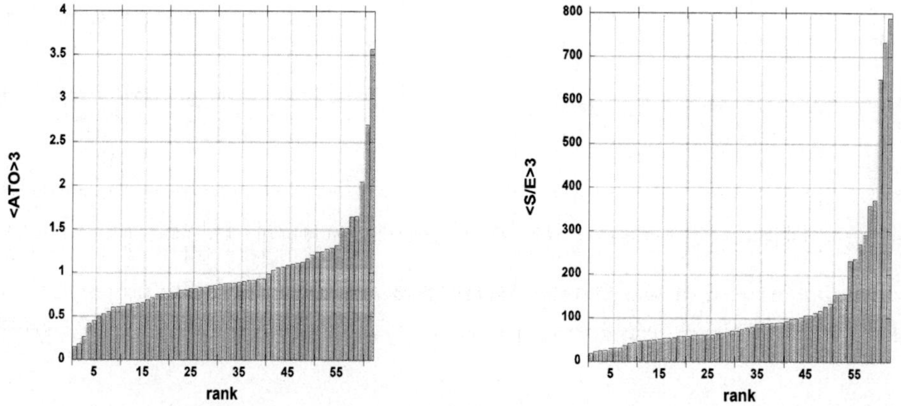


**Fig. 1.8** (colour online) Left panel: asset turnovers <ATO>3. Right panel: sales per employee <S/E>3; both ranked in increasing order—for the 62 SMEs, particularly pointing to a few relevant SMEs of the STAR market so studied

**Table 1.3** Summary of (rounded) statistical characteristics for the time average distributions of the innovation and performance indicators for the 62 STAR companies, in the centre of the table, in per cent and in $10^6$ Euros, respectively; the skewness and kurtosis are dimensionless scalars

| Variable | Min. | Max. | Sum | Mean ($\mu$) | StDev ($\sigma$) | Skewness | Kurtosis |
|---|---|---|---|---|---|---|---|
| <TIAX>2 | 174.5 | $1.192\ 10^5$ | $8.421\ 10^6$ | 13 583 | 22 513 | 2.7259 | 8.0364 |
| <TTA>2 | 86.5 | $5.075\ 10^5$ | $2.746\ 10^6$ | 44 297 | 92 600 | 3.3967 | 12.062 |
| <DS>3 | −0.1924 | 1.1767 | 4.9303 | 0.0795 | 0.198 | 3.1414 | 14.013 |
| <DA>3 | −0.1436 | 1.9818 | 7.8786 | 0.1270 | 0.330 | 3.8060 | 16.885 |
| <ROI>3 | −0.0768 | 0.3457 | 3.0115 | 0.0486 | 0.067 | 1.5342 | 5.1206 |
| <ROS>3 | −0.6609 | 0.2445 | 2.5316 | 0.0408 | 0.118 | −3.505 | 20.046 |
| <ATO>3 | 0.1474 | 3.5673 | 59.900 | 0.9661 | 0.535 | 2.4625 | 8.8557 |
| <S/E>3 | 17.464 | 787.69 | 7739.5 | 124.83 | 155.6 | 2.9856 | 8.7591 |

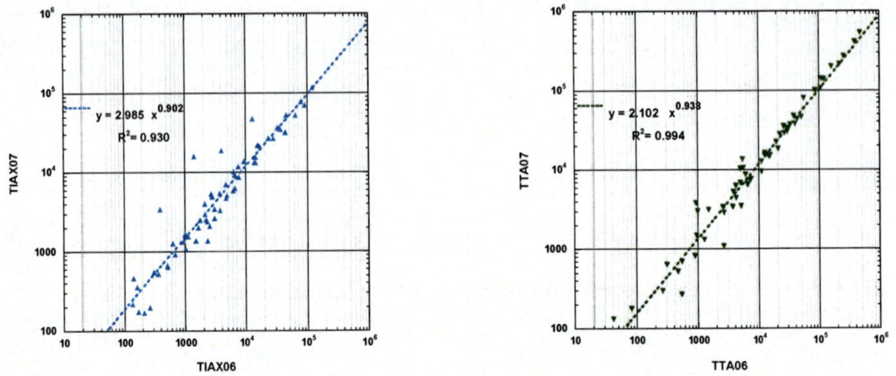


**Fig. 1.9** Power law regression analysis for (colour online) left panel: TIAX07 versus TIAX06 and right panel: TTA07 versus TTA06, for the 62 SMEs

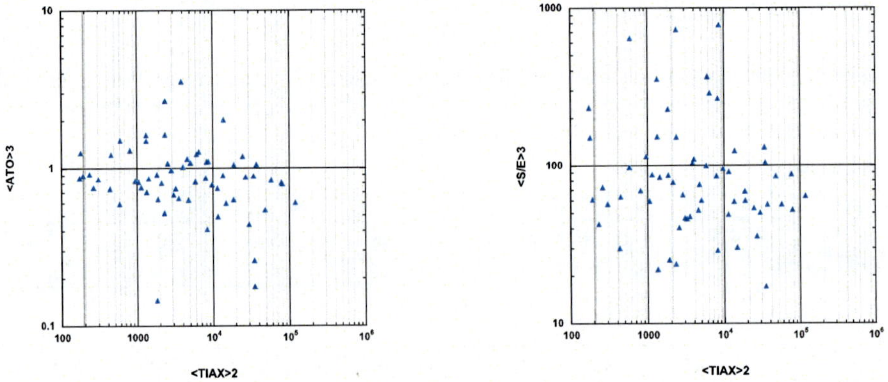


**Fig. 1.10** Searching for correlations: (colour online) left panel: <ATO>3 versus <TIAX>2; right panel: <S/E>3 versus <TIAX>2

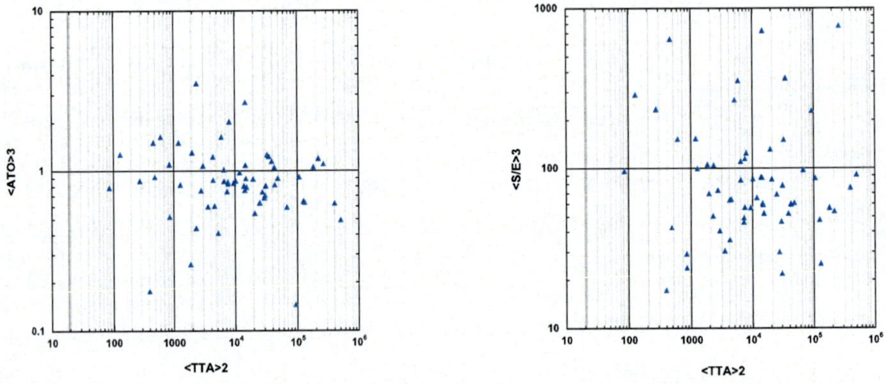


**Fig. 1.11** Searching for correlations: (colour online) left panel: <ATO>3 versus <TTA>2; right panel: <S/E>3 versus <TTA>2

<DA>3, <ROI>3 and <ROS>3; some SMEs have negative values in the latter cases; see Fig. 1.6 and on Fig. 1.7, for examples.

## *A Brief Description of the Voronoi Tessellation*

The Voronoi tessellation is a method for decomposing a metric space in non-overlapping subsets. Such a methodology dates back to René Descartes, who informally described it in his Principia Philosophiae (Descartes, 1644). Later, it was formalized in the context of the multidimensional real spaces (Voronoi, 1908). The principles behind the conceptualization of the Voronoi tessellation are grounded on the criterion used for decomposing the space. Some specific points—the so-called "centroids" or "seeds"—are initially selected. In our context, we refer to a finite number of centroids. Then, the space is partitioned into regions, according to the distances from the seeds. Specifically, each point of the space is assigned to the peculiar centroid which is closer to it. In so doing, the points assigned to a given centroid form a region which contains the centroid itself and does not overlap with other regions/centroids. When all the points of the space are assigned to a specific centroid, then the space appears visually as "tesselled"; this intuitively suggests why one refers to the "Voronoi tessellation". The distance employed for the tessellation procedure can be selected in a number of ways, and it is based on the metric. Here and in most applications—and also in the original Voronoi's paper—the considered metric space is the multidimensional Euclidean space. Thus, the natural Voronoi distance is the Euclidean one.

In the present application, we refer to bidimensional Euclidean spaces; the coordinates of the considered points and centroids are $x$- and $y$-variables.

## *Voronoi Correlations Approach*

In the context of Voronoi tessellation of the bidimensional Euclidean space, the $x$- and $y$-axes correspond thereafter to one of the innovation ($\mathcal{I}$) and one of the performance ($\mathcal{P}$) variables, respectively. It is easily understood that counting only correlations between innovation ($\mathcal{I}$) and performance ($\mathcal{P}$) variables, one has 12 displays; the more so if one considers the log of the variables for display readability ("scaling"), since as pointed out the absolute value of several ($\mathcal{P}$) variables has to be taken before log-scaling, leading to 20 Voronoi maps. This seems to be fine for completeness, but too much for illustrating the purpose and its pedagogical approach at this time. Thus, only a few cases are illustrated thereafter: Figs. 1.12 and 1.13, for the relationship between <TTA>2 and <ROI>3 or <ROS>3. For readability, the $x$- and $y$-axes differ (are flipped) depending on the figure panel. However, this allows to observe the size of the extreme regions, in which, in some sense, the whole market is divided.

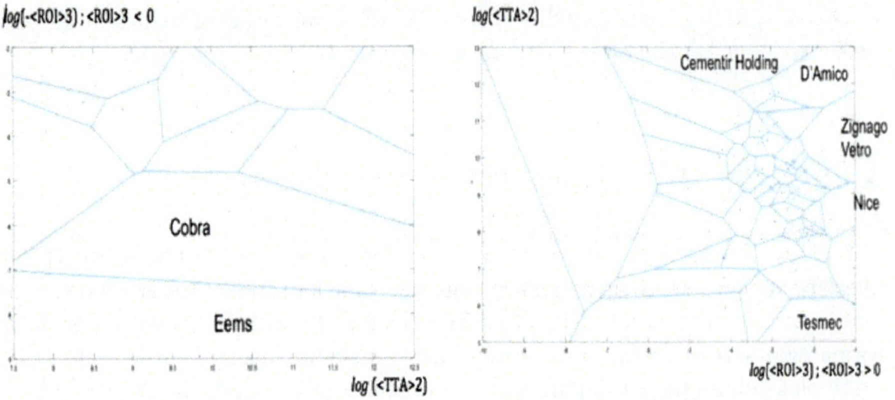


**Fig. 1.12** Voronoi tessellation of [log(<ROI>3), log(<TTA>2)] plane. Left panel: when <ROI>3 < 0. Right panel: when <ROI>3 > 0 for the 62 SMEs discussed in the text. A few specific SMEs are pointed out

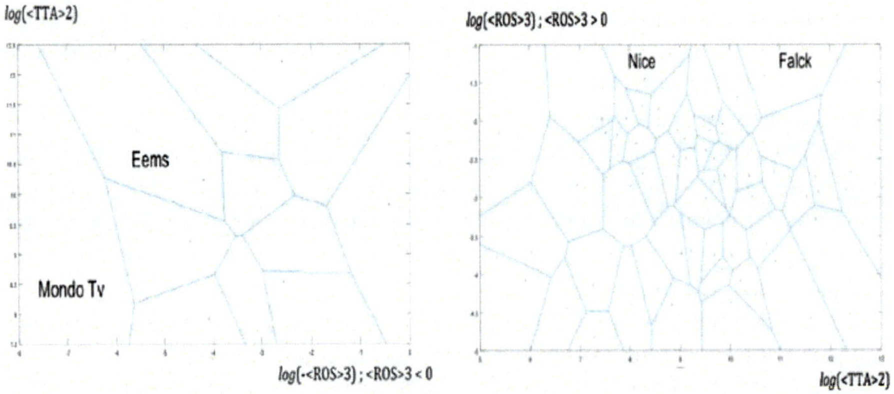


**Fig. 1.13** Voronoi tessellation of [log(<ROS>3), log(<TTA>2)] plane. Left panel: when <ROS>3 < 0; Right panel: when <ROS>3 > 0 for the 62 SMEs discussed in the text. A few specific SMEs are pointed out

## *Voronoi Clustering Approach*

In the Voronoi clustering approach, for avoiding scale effect, the variables of interest are first normalized. For each company $j = 1, \dots, 62$, we define

$$\bar{x}_j = \frac{x_j - m_x}{M_x - m_x}, \tag{1}$$

where $x_j$ represents the value of the variable $x$ for the $j$-th company and

$$m_x = \min_{j=1,\dots,62} x_j, \qquad M_x = \max_{j=1,\dots,62} x_j.$$

Next, in search of clusters, the centroids of the Voronoi tessellation are a priori defined by positive numbers, $\{\phi_h\}_{h=1}^{H}$ and $\{\psi_k\}_{k=1}^{K}$, where $H$ and $K$ are a priori chosen integers.

Moreover, we introduce a weighted Euclidean distance, for each innovation variable ($\mathcal{I}$), through

$$d_{\mathcal{I}}(j, \phi_h) = \sum_{x \in \mathcal{I}} \alpha_x (\bar{x}_j - \phi_h)^2, \tag{2}$$

for each centroid $\phi_h$ and where the $\alpha$'s are the non-negative weights of the norm, which can differ depending on the centroid, but so that

$$\sum_{x \in \mathcal{I}} \alpha_x = 1.$$

Analogously, for each performance variable ($\mathcal{P}$), we define

$$d_{\mathcal{P}}(j, \psi_k) = \sum_{x \in \mathcal{P}} \beta_x (\bar{x}_j - \psi_h)^2, \tag{3}$$

for each centroid $\psi_k$, imposing

$$\sum_{x \in \mathcal{P}} \beta_x = 1.$$

In so doing, all distances are $0 \le d_{\mathcal{I}}(j, \phi_h), d_{\mathcal{P}}(j, \psi_k) \le 1$, for each company $j$ with respect to centroid of coordinates $(\phi_h, \psi_k)$.

Notice that, even if Eqs. (2) and (3) look mathematically identical, we prefer to write down both formulas in order to point out that the differences may occur between the sets $\mathcal{I}$ and $\mathcal{P}$ and the related centroid coordinates. Indeed, as we will see below, the definition of the coordinates $\phi$'s and $\psi$'s and the different cardinalities of $\mathcal{I}$ and $\mathcal{P}$ lead to very different settings emphasized in the cases concerning Eqs. (2) and (3).

Three cases of clustering search have been examined in [3], always setting $H = K = 4$, with the centroids regularly distributed on the plane diagonal: $\{\phi_h\}_{h=1}^{H} = \{\psi_k\}_{k=1}^{K} = \{1/5, 2/5, 3/5, 4/5\}$. Consider case [$II$], for discussion, when $\alpha_x = 1/2$ for each $x \in \mathcal{I}$ and an identical weight for the $x \in \mathcal{P}$ variables, i.e. $\beta_x = 1/7$. This is a "uniform in value" case, where the definition of the centroids is made by considering a uniform decomposition of the interval [0, 1] and all the variables are assumed to equally concur in the Voronoi distance (Table 1.4).

It should be pointed out here that after some simple clustering analysis, so-called case [$I$], in [3], nine companies are controlling the clustering, and collapsing the whole sample into one single cluster, due to their "outlier aspect". They are (2) Aeffe, (5) Ascopiave, (15) Cementir Holding, (20) D'Amico, (22) Digital Bros, (30) Esprinet, (45) Mondo Tv, (58) Ternienergia, (59) Tesmec. This numerically confirms the few outlined cases seen in the above figures. These nine companies are removed

**Table 1.4** Distribution of companies among the clusters (*cl*.), either for clustering *II*, as defined in the text and examined in [3] for 53 STAR companies

| "II clustering" | | Performance | | | | |
|---|---|---|---|---|---|---|
| | | *1st cl.* | *2nd cl.* | *3rd cl.* | *4th cl.* | Tot. |
| Innovation | *1st cl.* | 16 | 22 | 7 | 0 | 45 |
| | *2nd cl.* | 2 | 2 | 0 | 0 | 4 |
| | *3rd cl.* | 1 | 3 | 0 | 0 | 4 |
| | *4th cl.* | 0 | 0 | 0 | 0 | 0 |
| Tot. | | 19 | 27 | 7 | 0 | 53 |

for the subsequent Voronoi clustering analysis approach. It remains, therefore, 53 companies to be examined.

To provide comments on the following results, we call *first* cluster the one associated with the $\{\phi_h\}_{h=1}^H = \{\psi_k\}_{k=1}^K = \{1/5\}$ centroid and, in an increasing way, the *second* and the *third* cluster, so that the *fourth* cluster is the one associated to the values $\{\phi_h\}_{h=1}^H = \{\psi_k\}_{k=1}^K = \{4/5\}$.

In Table 1.5, a description of the clusters of the sample companies is provided.

Referring to innovation clustering, the greatest number of companies (45 out of 53) is located in the first cluster, meaning that, in relative terms, companies undertake low-value innovation initiatives (at least those which produce reflections on tangible and intangible assets). Total sales and total assets, which are measures largely employed in literature for company size, show that the higher the intensity of innovation, the higher the size, or conversely. Also the incidence of both tangible and intangible assets (as percentage of the total assets) is increasing in the three innovation clusters, meaning that in highly innovative companies, tangible and intangible assets represent a relevant portion of the disclosed total assets. The mean/std. dev. ratio shows that the composition of the clusters is rather heterogeneous except for the third innovation cluster which is composed of companies whose size is fairly concentrated around the mean. For what concern performance, the distribution of companies among the clusters is quite different from that of innovation.

This provides evidence that the association between innovation and performance is not self-evident. The averages in the performance clusters also do not allow to appraise significant differences neither in terms of company size nor in terms of incidence of tangible and intangible assets.

Table 1.6 shows the averages of innovation and performance drivers for the entire sample and so-called clustering *II* analysis approach [3] for innovation and performance.

For completeness, we reproduce comments from such a publication. In the first cluster, the averages of innovation for tangible and intangible assets are below the general averages of the entire sample, whereas all performance indicators are above the full sample averages. In the second cluster, a general under-the-general-average

**Table 1.5** Statistical description of the companies, as if one full sample, or "belonging" to a cluster *cl.* (see text); the number (N) of companies in each cluster is given; *Tot.Ass.* = "Total Assets"

| | | | *Tot.Ass.* 2006–2007 (€/1,000) | Total Sales (€/1,000) | Intangible Assets on *Tot.Ass.* | Tangible Assets on *Tot.Ass.* |
|---|---|---|---|---|---|---|
| | N: | Mean | 303,053 | 267,689 | 5% | 10% |
| | 53 | Std.Dev. | 304,144 | 261,828 | 6% | 12% |
| "II clustering" Innovation | | | | | | |
| | | Mean | 241,199 | 210,452 | 4% | 8% |
| 1*st cl.* | 45 | Std.Dev. | 190,099 | 187,814 | 5% | 8% |
| | | Mean/St.Dev. | 1.27 | 1.12 | 0.88 | 1.04 |
| | | Mean | 731,745 | 467,442 | 8% | 16% |
| 2*nd cl.* | 4 | Std.Dev. | 798,924 | 469,023 | 8% | 22% |
| | | Mean/St.Dev. | 0.92 | 1.00 | 0.97 | 0.73 |
| | | Mean | 570,226 | 711,853 | 12% | 22% |
| 3*rd cl.* | 4 | Std.Dev. | 261,414 | 329,233 | 9% | 27% |
| | | Mean/St.Dev. | 2.18 | 2.16 | 1.39 | 0.82 |
| "II clustering" Performance | | | | | | |
| | | Mean | 210,607 | 157,375 | 8% | 10% |
| 1*st cl.* | 19 | Std. Dev. | 130,310 | 118,218 | 7% | 11% |
| | | Mean/St.Dev. | 1.62 | 1.33 | 1.14 | 0.94 |
| | | Mean | 385,628 | 354,434 | 3% | 9% |
| 2*nd cl.* | 27 | Std.Dev. | 387,615 | 293,834 | 3% | 13% |
| | | Mean/St.Dev. | 0.99 | 1.21 | 0.90 | 0.71 |
| | | Mean | 235,476 | 232,525 | 3% | 10% |
| 3*rd cl.* | 7 | Std.Dev. | 228,112 | 340,089 | 5% | 12% |
| | | Mean/St.Dev. | 1.03 | 0.68 | 0.62 | 0.85 |

performance is associated to an above-the-general-average innovation. In the third cluster, the performance averages are mixed: some of them are above the mean, while the others are below.

Specifically, the $\mu/\sigma$ ratio (the inverse of the coefficient of variation) points that the cluster's homogeneity is rather low, meaning that extremely different companies lie within the same cluster both from innovation or performance perspective. The only exception is represented by ATO, since the $\sigma$ is remarkably concentrated around the average $\mu$. This could be interpreted as a possible association between innovation and asset turnover. However, its direction remains unclear, since a high ATO is associated to a low innovation in the first cluster, whereas a low ATO is associated to a medium innovation in the second cluster, while high ATO is associated to high innovation in the third cluster.

**Table 1.6** Main statistical characteristics of the innovation and performance variables for the whole sample (Ent.) or inside the clusters (*cl.*); the number (N) of companies inside each cluster is recalled

| | | TIAX | TTA | DS | DA | ROI | ROS | ATO | S/E |
|---|---|---|---|---|---|---|---|---|---|
| Ent. | Mean ($\mu$) | 12,360 | 29,215 | 6% | 9% | 5% | 5% | 0.91 | 275.77 |
| | Std.Dev.($\sigma$) | 18,695 | 45,380 | 14% | 16% | 5% | 7% | 0.34 | 231.20 |
| N = 53 | $\mu/\sigma$ | 0.66 | 0.64 | 0.46 | 0.57 | 0.85 | 0.75 | 2.68 | 1.19 |
| Innovation II clustering | | | | | | | | | |
| *1st* | Mean ($\mu$) | 7,127 | 18,554 | 7% | 9% | 5% | 6% | 0.93 | 293.83 |
| *cl.* | Std.Dev.($\sigma$) | 9,311 | 24,023 | 15% | 17% | 6% | 7% | 0.36 | 248.53 |
| N=45 | $\mu/\sigma$ | 0.77 | 0.77 | 0.49 | 0.55 | 0.87 | 0.79 | 2.57 | 1.18 |
| *2nd* | Mean ($\mu$) | 28,081 | 71,512 | 4% | 3% | 2% | 2% | 0.67 | 165.90 |
| *cl.* | Std.Dev.($\sigma$) | 29,505 | 65,192 | 11% | 7% | 6% | 10% | 0.13 | 58.91 |
| N=4 | $\mu/\sigma$ | 0.95 | 1.10 | 0.34 | 0.35 | 0.25 | 0.16 | 5.36 | 2.81 |
| *3rd* | Mean ($\mu$) | 55,511 | 106,862 | 1% | 14% | 4% | 5% | 0.97 | 182.49 |
| *cl.* | Std.Dev.($\sigma$) | 28,454 | 107,418 | 5% | 18% | 1% | 2% | 0.20 | 48.84 |
| N=4 | $\mu/\sigma$ | 1.95 | 0.99 | 0.19 | 0.81 | 3.00 | 2.57 | 4.91 | 3.73 |
| Performance II clustering | | | | | | | | | |
| *1st* | Mean ($\mu$) | 16,356 | 22,484 | −2% | −4% | 0% | 0% | 0.79 | 222.75 |
| *cl.* | Std.Dev.($\sigma$) | 20,762 | 31,548 | 10% | 4% | 4% | 7% | 0.28 | 157.65 |
| N=19 | $\mu/\sigma$ | 0.79 | 0.71 | 0.21 | 0.83 | 0.05 | 0.05 | 2.85 | 1.41 |
| *2nd* | Mean ($\mu$) | 11,930 | 35,841 | 9% | 12% | 7% | 8% | 0.94 | 262.81 |
| *cl.* | Std.Dev.($\sigma$) | 19,346 | 56,259 | 9% | 13% | 4% | 5% | 0.25 | 232.07 |
| N=27 | $\mu/\sigma$ | 0.62 | 0.64 | 0.94 | 0.92 | 1.54 | 1.46 | 3.77 | 1.13 |
| *3rd* | Mean ($\mu$) | 3,174 | 21,931 | 20% | 35% | 9% | 12% | 1.11 | 469.70 |
| *cl.* | Std.Dev.($\sigma$) | 4,744 | 32,962 | 24% | 17% | 6% | 9% | 0.65 | 332.70 |
| N=7 | $\mu/\sigma$ | 0.67 | 0.67 | 0.84 | 2.02 | 1.49 | 1.26 | 1.71 | 1.41 |

It is worth noticing that in the third cluster, companies appear rather homogeneous in terms of performance, particularly for profitability (both ROI and ROS) and efficiency (ATO and S/E). One can argue that, above a particular "threshold of innovation intensity", the level performances seem rather homogeneous.

Indeed, similar considerations can be made for performance clustering: the performance averages gradually increase from the first to the third cluster, whereas innovation averages decrease (TIAX) or fluctuate (TTA). The relation innovation-performance seems, then, quite puzzling. Even for performance clustering, heterogeneity generally occurs within the clusters except for ATO.

## *Evidence from Outliers Approach*

Since there is a negative minimum for each DS, DA, ROI and ROS, one may guess that some board innovation strategies were rather failures.[3] One observes also some outliers from simple Voronoi analysis; see case "Clustering I" in [3]. From Table 1.2, one observes that the kurtosis is always positive and large, indicating lesser chances of extreme negative outcomes. The skewness is also positive, indicating a long upper tail (many small losses and a few extreme gains) and a long lower range tail (many small gains and several extreme losses).

The performance efficiency ratios of the (62) companies, taken one by one, one observes several outliers, i.e. when the SME efficiency value falls outside the relevant $]\mu - 2\sigma, \mu + 2\sigma[$ interval. There are three SMEs which are positive outliers: (58) Terrienergia, (11) Buongiorno, (13) Cairo Communications, and 1 SME which is systematically a "negative outlier": (45) Mondo TV, confirming the Voronoi analysis of "Clustering I".

It should occur to the reader that those four companies are those with very low TTA. Moreover, Mondo TV is the only one among the outliers which has a TTA06 lower than its TTA07—this SME had about a 50% decrease in investment before the crisis. In contrast, Terrienergia, Buongiorno and Cairo Communications have a relatively high TTA increase.

One can observe, respectively, from Figs. 1.12 and 1.13, for example, see also Figs. 1.5 and 1.6 in [4].

- the relationship between <ROI>3 and <TTA>2: a weak <ROI>3 for Cementir Holding and Ascopiave; a negative but with a large absolute value occurs for D'Amico; in contrast, a large <ROI>3 occurs for Tesmec, while the negatively largest <ROI>3 is for Eems—both firms have a rather low <TTA>2;
- the relationship between <ROS>3 and <TTA>2 indicates a moderate <ROS>3 positive effect for Sogefi, Ascopiave, D'Amico and Cementir Holding, the four largest TTA companies, "imposing" a single cluster in the "Clustering I" Voronoi analysis; a large negative <ROS>3 effect occurs for Mondo TV; on the other side, the most positive <ROS>3 is for Falck Renewables, Zignago Vetro and Nice.

Observe that these companies cover various sectors of activity. Nevertheless, there are differences: Terrienergia and Cairo Communications have very dissimilar performance efficiency behaviours, the former performing better for "growth", the latter for "profitability". Since Terrienergia, Buongiorno and Cairo Communications have a high increase in TTA, one might reach some advice concerning innovation strategy. Let TTA increase for better performance.

As an a "posteriori" analysis "proof", observe that Mondo TV did not increase its TTA, TTA07<TTA06, pointing to a deficient strategy. This is pointing to the timing of "investment" relevance—not fully clear from the Voronoi analysis. Moreover, observe that the average values are not the critical quantities.

[3]In fact, it is not absolutely sure that innovation strategies were "failures", since one has no definite proof that innovations have been truly implemented.

## Conclusions

Justifying an investment is superb challenge for board members—the more so at financial crisis time. Payback is unsure; one needs criteria for obtaining efficiency (performance) measures, whence modelling strategies and forecasting. Usually one demands that the level of investments be low and the returns be high. We have here proposed a set of measures for such "research questions". We have outlined means for finding correlations and looked for clustering of performance ratios, whence the specific companies "obeying" strategies based on such ratios. But from the Voronoi clustering approach, it is not obvious that high innovation leads to high performance.

We have found that the timing of investment is very relevant from observing outliers, with either positive or negative results. Extreme values show best strategies!

The Voronoi approach is nevertheless of interest: Within the clusters, one can compare the characteristics and performance of companies holding the same innovation model, whereas between the clusters high heterogeneity should occur assuming that different innovation models might be suitable for different company profiles and/or could be associated to different level of performance.

In this respect, cluster analysis seems to be particularly effective in providing a global analysis of the relationship between innovation and performance. Notice that the study allows three considerations from extreme value analysis: not only the investment evolution—up or down, low or high—but also through their average, serving as a control kind of test. It should be obvious that the best performance should be better appreciated when (unexpectedly?) the investment is low, but regularly implemented.

## References

1. L. Agostini, A. Nosella, B. Soranzo, The impact of formal and informal appropriability regimes on SME profitability in medium high-tech industries. Technol. Anal. Strateg. Manag. **27**(4), 405–419 (2015)
2. D. Archibugi, Pavitt's taxonomy sixteen years on: a review article, economics of innovation and new technology. Econ. Innov. New Technol. **10**(5), 415–425 (2001)
3. M. Ausloos, F. Bartolacci, N.G. Castellano, R. Cerqueti, Exploring how innovation strategies at time of crisis influence performance: a cluster analysis perspective. Technol. Anal. Strateg. Manag. **30**(4), 484–497 (2018)
4. M. Ausloos, R. Cerqueti, F. Bartolacci, N.G. Castellano, Outliers for assessing how to optimize performance, SME investment best strategies. Phys. A **509**, 754–765 (2018)
5. D. Archibugi, A. Filippetti, M. Frenz, Economic crisis and innovation: is destruction prevailing over accumulation? Res. Policy **42**(2), 303–314 (2013)
6. F. Bartolacci, N.G. Castellano, R. Cerqueti, The impact of innovation on companies'performance: an entropy-based analysis of the STAR Market Segment of Italian Stock Exchange. Technol. Anal. Strateg. Manag. **27**(1), 102–123 (2015)
7. C.F. Baum, H. Lööf, P. Nabavi, A. Stephan, A new approach to estimation of the *R& D*-innovation-productivity relationship. Econ. Innov. New Technol. **26**(1/2), 121–133 (2017)
8. S. Brusoni, E. Cefis, L. Orsenigo, Innovate or Die? A critical review of the literature on innovation and performance, CESPRI, Centro di Ricerca sui Processi di Innovazione e Internazionalizzazione, Universitá Commerciale Luigi Bocconi, Working Paper n. 179 (2006)

9. E.G. Ceptureanu, S.I. Ceptureanu, C.E. Luchian, I. Luchian. Quality management in project management consulting. A case study in an international consulting company. Amfiteatru Econ. **19**(44), 215–230 (2017)
10. S. Cesaratto, S. Mangano, Technological profiles and economic performance in the Italian manufacturing sector. Econ. Innov. New Technol. **2**, 237–256 (1993)
11. D. Chun, Y. Chung, S. Bang, Impact of firm size and industry type on *R* & *D* efficiency throughout innovation and commercialisation stages: evidence from Korean manufacturing firms. Technol. Anal. Strateg. Manag. **27**(8), 895–909 (2015)
12. R.G. Cooper, The strategy-performance link in product innovation. R & D Manag. **14**(4), 247–259 (1984)
13. J.P.J. De Jong, O. Marsili, The fruit flies of innovations: A taxonomy of innovative small firms. Res. Policy **35**(2), 213–229 (2006)
14. R. Descartes, Principia Philosophiae. Ludovicus Elzevirius, Amsterdam (1644)
15. C. Duyckaerts, G. Godefroy, Voronoi tessellation to study the numerical density and the spatial distribution of neurones. J. Chem. Neuroanatomy **20**(1), 83–92 (2000)
16. L. Dwyer, R. Mellor, Product innovation strategies and performance of Australian firms. Aust. J. Manag. **18**(2), 159–180 (1993)
17. A. Filippetti, D. Archibugi, Innovation in times of crisis: National Systems of Innovation, structure, and demand. Res. Policy **40**(2), 179–192 (2011)
18. A. Gadomski, N. Kruszewska, On clean grain-boundaries involving growth of nonequilibrium crystalline-amorphous superconducting materials addressed by a phenomenological viewpoint. Eur. Phys. J. B **85**(12), 1–8 (2012)
19. M. Gligor, M. Ausloos, Cluster structure of EU-15 countries derived from the correlation matrix analysis of macroeconomic index fluctuations. Eur. Phys. J. B **57**, 139–146 (2007)
20. M. Gligor, M. Ausloos, Convergence and cluster structures in EU area according to fluctuations in macroeconomic indices. J. Econ. Integr. **23**, 297–330 (2008a)
21. M. Gligor, M. Ausloos, Clusters in weighted macroeconomic networks: the EU case Introducing the overlapping index of GDP/capita fluctuation correlations. Eur. Phys. J. B **63**, 533–539 (2008b)
22. I. Gocer, S. Alatas, O. Peker, Effects of *R* & *D* and innovation on income in EU countries: new generation panel cointegration and causality analysis. Theor. Appl. Econ. **23**(4), 153–164 (2016)
23. A. Heirman, B. Clarysse, Which tangible and intangible assets matter for innovation speed in start-ups? J. Prod. Innov. Manag. **24**(4), 303–315 (2007)
24. M.A. Hitt, R.E. Hoskisson, H. Kim, International diversification: effects on innovation and firm performance in product diversified firms. Acad. Manag. J. **40**(4), 767–798 (1997)
25. H. Hollenstein, Innovation modes in the Swiss service sector: a cluster analysis based on firm-level data. Res. Policy **32**, 845–863 (2003)
26. M.B. Jensen, B. Johnson, E. Lorenz, B.A. Lundvall, Forms of knowledge and modes of innovation. Res. Policy **36**, 680–693 (2007)
27. A.M. Khan, V. Manopichetwattana, Innovative and non-innovative small firms: types and characteristics. Manag. Sci. **15**(5), 597–606 (1989)
28. E. Kirner, S. Kinkel, A. Jaeger, Innovation paths and the innovation performance of low-technology firms - An empirical analysis of German industry. Res. Policy **38**, 447–458 (2009)
29. S. Latham, M. Braun, Economic recessions, strategy, and performance: a synthesis. J. Strateg. Manag. **4**(2), 96–115 (2011)
30. M.W. Lawless, P.C. Anderson, Generational technological change: effects of innovation and local rivalry on performance. Acad. Manag. J. **39**(5), 1185–1217 (1996)
31. A. Leiponen, I. Drejer, What exactly are technological regimes? Intra-industry heterogeneity in the organization of innovation activities. Research Policy **36**, 1221–1238 (2007)
32. X.T. Liu, X.Q. Zheng, D.B. Li, Voronoi Diagram-Based Research on Spatial Distribution Characteristics of Rural Settlements and Its Affecting Factors. A Case Study of Changping District, Beijing [J]. J. Ecol. Rural Environ. **2**, 007 (2009)

33. S. Montresor, A. Vezzani, Intangible investments and innovation propensity: Evidence from the Innobarometer 2013. Ind. Innov. **23**(4), 331–352 (2016)
34. S. Nunes, R. Lopes, Firm performance, innovation modes and territorial embeddedness. Eur. Plann. Stud. **23**(9), 1796–1826 (2015)
35. OECD, *Annual Report* (OECD Publishing, Paris, 2005)
36. OECD, *Annual Report* (OECD Publishing, Paris, 2009)
37. N.K. Park, U.D. Park, J. Lee, Do the performances of innovative firms differ depending on market-oriented or technology-oriented strategies? Ind. Innov. **19**(5), 391–414 (2012)
38. K. Pavitt, Sectoral patterns of technical change: towards a taxonomy and a theory. Res. Policy **13**, 343–373 (1984)
39. M. Ramella, W. Boschin, D. Fadda, M. Nonino, Finding galaxy clusters using Voronoi tessellations. Astron. Astrophys. **368**(3), 776–786 (2001)
40. M. Ranga, H. Etzkowitz, Great expectations: an innovation solution to the contemporary economic crisis. Eur. Plann. Stud. **20**(9), 1429–1438 (2012)
41. A. Renzi, C. Simone, Innovation, tangible and intangible resources: the espace of slacks interaction. Strateg. Change **20**(1–2), 59–71 (2011)
42. N. Shin, K.L. Kraemer, J. Dedrick, R&D and firm performance in the semiconductor industry. Ind. Innov. **24**(3), 280–297 (2017)
43. M. Srholec, B. Verspagen, The voyage of the beagle into innovation: explorations on heterogeneity, selection, and sectors. Ind. Corporate Change **21**(5), 1221–1253 (2012)
44. A. Sterlacchini, F. Venturini, R&D and Productivity in High-Tech Manufacturing: A Comparison between Italy and Spain. Ind. Innov. **21**(5), 359–379 (2014)
45. C.Y. Tseng, H.-Y. Hui-Yueh Kuo, S.S. Chou, Configuration of innovation and performance in the service industry: evidence from the Taiwanese hotel industry. Ser. Ind. J. **28**(7), 1015–1028 (2008)
46. E. Vaz, T. de Noronha Vaz, P.V. Galindo, P. Nijkamp, Modelling innovation support systems for regional development analysis of cluster structures in innovation in Portugal. Entrepren. Region. Dev. **26**(1–2), 23–46 (2014)
47. G.F. Voronoi, Nouvelles applications des paramétres continus de la théorie de formes quadratiques. Journal für die reine und angewandte Mathematik **134**, 198–287 (1908)
48. W.F. Yushimito, M. Jaller, S. Ukkusuri, A Voronoi-based heuristic algorithm for locating distribution centers in disasters. Netw. Spatial Econ. **12**(1), 21–39 (2012)
49. S.A. Zahra, J.G. Covin, Domestic and international competitive focus, technology strategy and company performance: an empirical analysis. Technol. Anal. Strateg. Manag. **6**(1), 39–54 (1994)

# Chapter 2
# Enhancing Maximum Likelihood Estimation of Infection Source Localization

**Robert Paluch, Łukasz Gajewski, Krzysztof Suchecki, Bolesław Szymański, and Janusz A. Hołyst**

**Abstract** The studies on the spreading processes over complex networks have increasingly wide applications. Sometimes this is very important to find quickly the origin of spread, e.g. the patient zero in epidemics or the origin of fake news forwarded in social network. The method of maximum likelihood estimation proposed by Pinto et al. (PTV) solves the important case of this problem in which a limited set of nodes act as observers and report times at which the spread has reached them. While the PTV algorithm has been shown to be optimal on trees there are several challenges remaining on general graphs. One important issue is the complexity $O(N^\alpha)$ where $N$ is the size of the network and $\alpha \in (3, 4)$ depending on the network topology and the number of observers. We address this issue with a new approach in which observers with low-quality information (i.e. with large spread encounter times) are ignored and potential sources are selected based on the likelihood gradient from high-quality observers. Our gradient maximum likelihood algorithm (GMLA) reduces this complexity to $O(N^2 log(N))$. The other issue we address here is of precision on general graphs. The original PTV approach assumes the information travels via a single, shortest path, which by this assumption is supposed to be the fastest way. We show that such assumption leads to the overestimation of propagation time in networks where multiple potential traversal paths exist. We propose a new method of source estimation based on maximum likelihood principle that takes into account the existence of multiple shortest paths. We test our modifications extensively on both synthetic and real networks and show that they successfully address the afore-

R. Paluch (✉) · Ł. Gajewski · K. Suchecki · J. A. Hołyst (✉)
Faculty of Physics, Warsaw University of Technology, Koszykowa 75, 00-662 Warsaw, Poland
e-mail: jholyst.fens@gmail.com

B. Szymański
Social Cognitive Networks Academic Research Center, Rensselaer Polytechnic Institute, 110 8th Street, Troy, NY 12180-3590, USA

Społeczna Akademia Nauk, Henryka Sienkiewicza 9, 90-113 Łódź, Poland

J. A. Hołyst
ITMO University, 49 Kronverkskiy av., 197101 Saint Petersburg, Russia



D. Grech and J. Miśkiewicz (eds.), *Simplicity of Complexity in Economic and Social Systems*, Springer Proceedings in Complexity,
https://doi.org/10.1007/978-3-030-56160-4_2

mentioned issues and thus enhancing the PTV method of locating the source of a spread on complex networks.

## Introduction

Our society is built of many types of interacting networks which collect, process and transmit a plethora of information in every moment [1–4]. Growing connectedness of the world increases the risk of global pandemics, viruses both physical and virtual, as well as misinformation and rumours with often serious effects [5–8]. As an example, a fake tweet about explosion in the White House caused $130 billion loss on the stock market in 2013 [9]. Sometimes rumours or fake news become viral on social media and might affect people choices as in 2016 during the United States presidential election [10]. Developing effective tools for detecting and suppressing spread of dangerous viruses or fake news is a great challenge for network and data scientists. In order to stop the spread of a malicious content it is necessary to localize its origin. The answer of Shah and Zaman for this problem was *rumour centrality* [11], which is the number of distinct ways a rumour can spread in the network starting from a given node. They proved that the node with maximum *rumour centrality* is the maximum likelihood estimator for the origin of spread if the underlying graph is a regular tree. This method requires full knowledge about all the links between nodes and the infection status of all nodes at a given moment. Pinto, Thiran and Vetterli [12] relaxed some of these constraints since their algorithm does not need information about state of every node, but only about some fraction of chosen nodes called *observers*. After these two publications, the topic of the source detection became popular and many other variants of this problem have been studied. These two works popularized the topic of the source localization and gave rise to many other investigations. Two main approaches can be distinguished to this problem: the snapshot-based [11, 13–15] and the detectors-based [12, 16, 17] source localization. The snapshot-based methods need the information about nodal states of an entire network (at a whichever time moment after the outbreak), while the detectors-based algorithms have to monitor only a small subset of nodes but all the time. Regardless of the above division, researchers also considered different epidemic models [15, 18], spreading at weighted or time-varying graphs [19–22] and multi-source detection problems [23, 24]. Despite the interest of the scientific community, Jiang et al. concluded in their state-of-the-art paper [25] that current methods are too slow to be used for a quick identification of the propagation source in large complex networks. Another challenge with the method presented in [12] is the assumption of a tree-like propagation which is often not satisfied. The signal always travels via the *fastest* route. On trees it is equivalent to the topologically *shortest* path but this is not true for general graphs. Both the computational complexity and multiple possible propagation paths have been addressed by us before in [26, 27], in which we also presented the results on both real and synthetic networks. The chapter compares and combines the approaches and shows how to modify the maximum likelihood estimation approach in various scenarios in order to obtain satisfying results in terms of increase in speed or precision.

The organization of this paper is as follows. In the second section, we discuss Pinto-Thiran-Vetterli algorithm which is the method of maximum likelihood estimation. In the third section, we introduce our gradient maximum likelihood algorithm (GMLA) and compare it with PTV approach. The fourth section is addressed to the issue of correlation between propagation paths, while in the fifth section we present possible solutions for this problem. The sixth section contains conclusions and the summary of our work.

## Pinto-Thiran-Vetterli Algorithm

We assume a general undirected graph that can model a plethora of systems from transportation to social media. Within this graph we choose a node that will function as a source of a signal. Said node initiates the spread via sending an information to its adjacent nodes. Each subsequent "infected" vertex now begins spreading this information themselves. The process comes to a halt when all nodes have received the message. Such a process can be modelled in several ways, most notably with an epidemic SI model where each node has a certain probability per time step of infecting their neighbour. Alternatively, we can use the framework of assigning real-valued time delays on each edge of the graph. Those delays are random values from a known and arbitrary distribution. We use both of these methods and in the latter we assume the delays to be normally distributed. Both the topology of the system and the distribution (but not the specific values on edges) are known. Several nodes, chosen at random, are what we call observers—they provide us with the exact time they received the spreading signal. What is unknown and desired to be found is the source of the spread. We stem our methods from an algorithm proposed by Pinto et al. (referred to as PTV in this paper) that is a maximum likelihood estimation (MLE) approach. To reiterate we need the full picture of the network topology $\mathcal{G}$ and of times $t_k$, $1 \leqslant k \leqslant K$, at which the information arrived at observer $k$ (*observed delays*) as it is required by MLE method. One of the observers, we shall denote it as $o_K$ is a so-called *reference observer*. As we do not know the initial time of spread it is necessary to perform computations in a reference frame. In PTV there is an underlying assumption that the signal travels through the system along the shortest paths and the propagation times $\theta_i$ for each edge are independently and identically distributed (i.i.d) normal random variables, for which the mean $\mu$ and the variance $\sigma^2$ are known. In the case of a general graph (in contrast to a tree), the actual propagation tree (which is not known a priori) is approximated by a breadth-first search (BFS) tree composed of the shortest paths between the node which is supposed to be the source and all observers. The algorithm boils down to computing a score for each node $s$ using a density function of $(K-1)$-dimensional normal distribution, where

$K$ is the number of observers. A vertex that maximizes said score is the one we classify as the source. Let $|\mathcal{P}(u, v)|$ denote the length of the path $\mathcal{P}(u, v)$ connecting two nodes $u$ and $v$, $\mathbf{d}$ be the vector of observed delays, $\boldsymbol{\mu}_s$ the vector of deterministic delays and $\boldsymbol{\Lambda}_s$ the delay covariance matrix. Then the aforementioned score of a given vertex $s$ is given as

$$\phi(s) = \frac{\exp\left(-\frac{1}{2}(\mathbf{d} - \boldsymbol{\mu}_s)^{\mathrm{T}}\boldsymbol{\Lambda}_s^{-1}(\mathbf{d} - \boldsymbol{\mu}_s)\right)}{|\boldsymbol{\Lambda}_s|^{1/2}}, \tag{1}$$

where

$$[\mathbf{d}]_k = t_k - t_K, \tag{2}$$

$$[\boldsymbol{\mu}_s]_k = \mu(|\mathcal{P}(s, o_k)| - |\mathcal{P}(s, o_K)|), \tag{3}$$

$$[\boldsymbol{\Lambda}_s]_{k,i} = \sigma^2|\mathcal{P}(o_K, o_k) \cap \mathcal{P}(o_K, o_i)|, \tag{4}$$

for $k, i = 1, \ldots, K - 1$. Despite the simplifications, the method works reasonably well under certain conditions (density of observers $\rho = K/N > 10\%$ and propagation ratio $\lambda = \mu/\sigma \geq \sqrt{2}$).

Note: In the original paper, Pinto et al. additionally use the information from whom given observer received the signal. We do *not* use this information as it is not always available in the real-world scenarios. Therefore from now on when we refer to PTV we mean a *limited information* variant of it.

PTV uses the whole network for inferring the source, which makes the algorithm computationally expensive. It calculates the score for each node in the network which requires to create BFS tree $N$ times. Each BFS tree covers practically the entire network since it reaches to all observers, including these which are far from the root node. The complexity of creating BFS tree is $O(N^2)$ per node $s \in \mathcal{G}$. Unfortunately, the assumption of diffusion along the shortest paths works well only for the observers which are close to the suspected node (the root of BFS tree) since the typical random networks are locally tree-like. However, this may not be true for observers which are far from the supposed source. Thus making use of the distant observers may be problematic—it increases the expenses and decreases the correctness (because there can be many possible paths between the source and the remote observers). So far we have considered the situation when the number of observers is much smaller than the number of other nodes. In this case, the overall complexity of PTV is $O(N)$ for arbitrary trees and $O(N^3)$ for general graphs. But what if the number of the observers is not constant and small? In order to keep quality of the results of PTV constant while increasing the size of the network, the number of observers also should increase. Assuming constant density of observers $\rho = K/N$, the complexity increases to $O(N^2)$ for arbitrary trees and $O(N^4)$ for general graphs (in the worst case), which makes the algorithm prohibitively computationally expensive when applied to large networks. As numerical results show (Fig. 2.1), when $K$ is increasing, matrix operations in (1) (inverting the matrix and obtaining its determinant) are more computationally demanding than creating BFS tree since they have time complexity

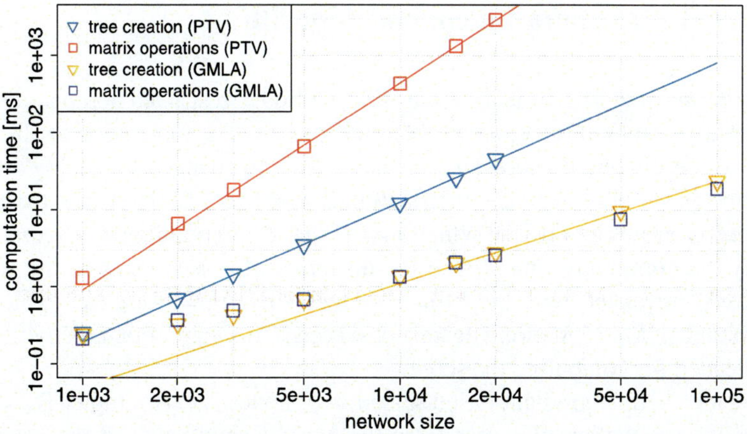


**Fig. 2.1** The comparison of the times needed for creation of the propagation tree and for matrix operations. The time complexity of matrix operation for PTV is $O(N^{2.68})$ per node (red line), while the time complexity of tree creation is $O(N^{1.81})$ per node for PTV (green) and $O(N^{1.34})$ for GMLA (blue). The tests were performed on BA network with $m = 3$. The observers were distributed randomly over whole network with density $\rho = 0.1$

$O(K^3)$ in the worst case. In general, algorithm's complexity for arbitrary graphs is $O(N(K^3 + N^2))$. Assuming $K \sim N^\gamma$, PTV complexity ranges from $O(N^3)$ when $\gamma \leqslant 2/3$ to $O(N^4)$ when $\gamma = 1$.

In Fig. 2.10, we compare an expected time of signal's traversal as per the PTV approach with a real time. Real as in experimentally obtained via a simulation. It is quite apparent that they are far from matching. There are two reasons for this clear discrepancy between the two. Firstly, PTV assumes the signal travels via a shortest path obtained with a BFS method; however, there can be more than one shortest path (and usually this is the case in non-tree networks). Since the signal propagates through the *fastest* path, every additional parallel trajectory provides the spread with more chances (as each edge has a random delay) for the information to arrive more quickly. Thus, instead of a distribution of traversal time from a single path, we deal with a distribution of a *minimum* from several different trajectories. Moreover, one has to account for the existence of independent and longer than shortest paths as while less likely they still can become faster than the topologically shortest paths. Therefore, the existence of such additional opportunities for a quick spread (even if not as likely) still contributes significantly to a decrease of the mean traversal time. The effect of longer paths being faster is the greater the smaller the propagation ratio $\mu/\sigma$. If said ratio is sufficiently large, the primary mechanism is that of multiple shortest paths.

As such we have dedicated ourselves to this very problem of the two and developed a new method based on the maximum likelihood that focuses on the existence of multiple shortest paths.

## Gradient Maximum Likelihood Algorithm

We propose two enhancements to the PTV method: a restricted number of observers and a gradient-based selection of potential sources. The motivation for limiting the number of observers is following: the value of score changes noticeably only with a first few tens of observers, and then stabilizes (see Fig. 2.2). However, the algorithm complexity scales as $O(K^3)$, which means that every additional observer increase greatly the computing time. Since the distances of observers from the source are positively correlated with their times of receiving signal, one can exploit only a small number $K_0 \ll K$ of the nearest observers. This correction greatly reduces the time needed for computing the score.

The idea behind gradient-based selection of nodes is very simple. The algorithm begins the calculation of scores from the nearest neighbours of the *observer one* (with the smallest reported time), because it is very likely that the origin of spread is in close proximity to this observer. Then, it picks the neighbour with the highest score and checks the scores for his nearest neighbours in order to find the one which has a score greater than or equal to the current maximum. The procedure is continued until it finds the node with the local highest score (see Fig. 2.3). The algorithm saves all computed scores to avoid double calculation and creates the ranking of nodes.

Numerical simulations show that the number of suspected nodes $N_0 = |V_s|$ increases linearly with the average degree $\langle k \rangle$ and logarithmically with the number of nodes in network $N_0 \sim \langle k \rangle \log(N)$ (see Fig. 2.4). It is worth noting that the algorithm does not guarantee that the true source $s^*$ will be selected for score calcu-

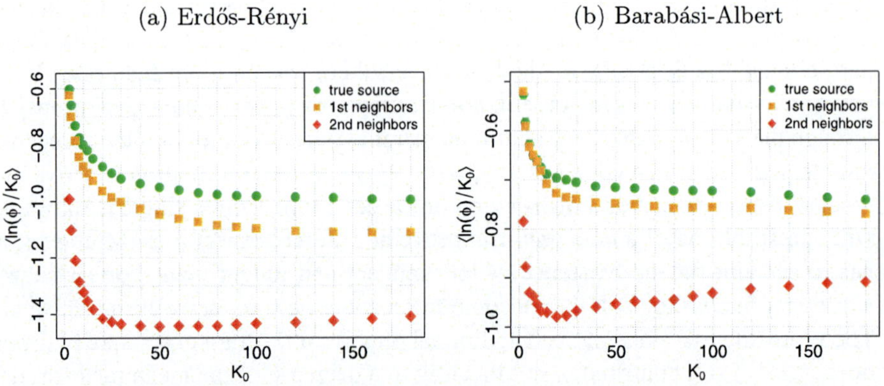


**Fig. 2.2** Normalized logarithm of the score versus the number of nearest observers which were used for the computation. The score of a node (the true source or its first neighbour or its second-order neighbour) was computed for various values of the number of the nearest observers. The results show that increasing the number of the nearest observers above a certain threshold does not change the score substantially. The tests were performed on **a** ER graph with $\langle k \rangle = 6$, **b** BA network with $m = 3$. The size of networks is $N = 1000$. The observers were distributed randomly over whole network with density $\rho = 0.2$. SI model was used for propagation with infection rates $\beta = 0.5$. The results are averaged over 100 realizations. The standard errors are smaller than symbols on plot

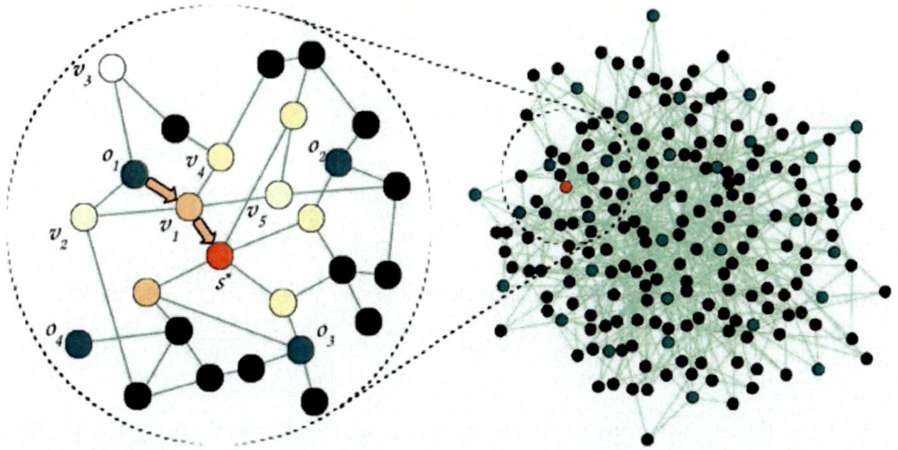


**Fig. 2.3** The visualization of GMLA. The right picture presents the whole graph. The red node is the true source. Dark green nodes mark the $K_0$ nearest observers with the smallest time delays (in this plot $K_0 = 4$). The rest of the observers in the network are highlighted in light green. The picture on the left is a zoom of a small area around the nearest observers. In this picture, the colour corresponds to the score (the likelihood of being the source) of the node (except for the observers which are dark green). The higher the score of the node is, the darker red is its colour. At the beginning the algorithm computes the scores for the neighbours of the nearest observer (in this plot *the observer one* is $o_1$ and its neighbours are $v_1$,$v_2$ and $v_3$). Afterwards GMLA selects the neighbour with the highest score ($v_1$ in this case) and starts computing the scores for its neighbours ($o_1$,$v_4$,$v_5$ and $s^*$). During this step there is no need for estimating the likelihood for the node $v_2$ since it was done in the previous step. All the scores which are computed are stored in the list. Since $s^*$ has the highest score among the neighbours of $v_1$, in the next step GMLA will compute the scores for it neighbours. None of the neighbours of $s^*$ has higher score than $s^*$, and therefore the algorithm stops here. The node $s^*$ is the source according to GMLA because it has the highest score from all tested (suspected) nodes. The nodes not visited by the algorithm are black since their scores are undefined (their scores are not computed)

lation, i.e. $P(s^* \in V_s) < 1$. Using the symbols $K_0$ and $N_0$ we write the time complexity of GMLA as $O(N_0(K_0^3 + N^2))$ in the worst case. Assuming $N_0 \sim \log(N)$ and $K_0 \ll N$, which is true for our method, the complexity can be further simplified into $O(N^2 \log(N))$. The limited number of observers was introduced earlier in works [12, 28], where the search algorithm was run before all $K$ observers get infected, to imitate more realistic scenario. This research pays attention on the improvement of algorithm's complexity for large complex networks.

The number of the nearest observers $K_0$ has a strong impact on GMLA and therefore should be carefully fine-tuned. Too small value of $K_0$ decreases the precision of the algorithm, while unnecessarily large $K_0$ increases the time of computation. We denote $K_0^*$ as the optimal number of the nearest observers, which is the minimum number of the nearest observers $K_0$ needed to get maximum quality of the source localization. We study how the network size, the average degree and the propagation ratio affects the value of $K_0^*$ for Erdős-Rényi (ER) and Barabási-Albert [29] (BA) networks. We do not observe substantial relationship between $K_0^*$ and the average

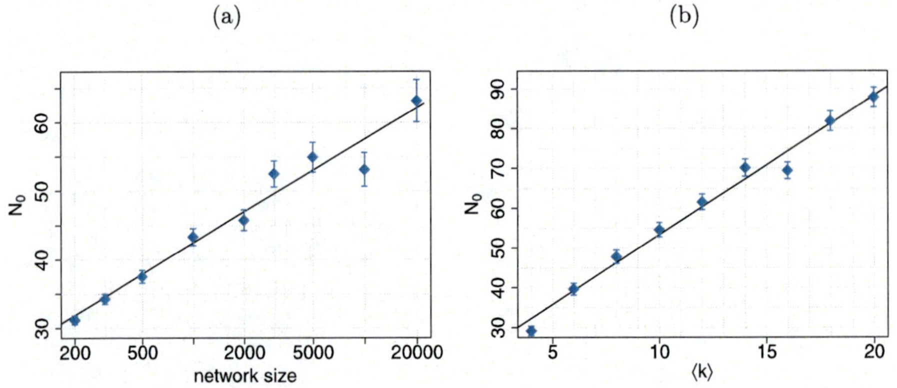


**Fig. 2.4** **a** The number of suspected nodes versus the size of BA network. The fit is a logarithmic function $N_0 = 6.58\ln(N) - 3$. Parameters: $\langle k\rangle = 6$, $\rho = 0.2$, $\beta = 0.5$, $K_0 = 0.5\sqrt{N}$, 1000 realizations. **b** The number of suspected nodes versus the average degree of BA network. The fit is a linear function $N_0 = 3.5\langle k\rangle + 18.2$. Parameters: $N = 2000$, $\rho = 0.2$, $\beta = 0.5$, $K_0 = 20$, 1000 realizations

degree of the network or the propagation ratio. Figure 2.5 shows the performance of GMLA versus the number of the nearest observers $K_0$ for various sizes of BA network with the minimum degree $m = 3$ ($m$ is the initial degree of each attached node, thus $\langle k\rangle = 2m = 6$). The peak of precision is visible for each size of BA network on Fig. 2.5a. Figure 2.5b presents the estimates of $K_0^*$ for different sizes of BA network and nonlinear least squares fit. No peak of precision but the saturation point is observed for Erdős-Rényi graph (see Supplemental Material for [26]). The occurrence of the precision peak for BA network (not only the saturation point like for ER graph) means that taking only $K_0 \ll N$ nearest observers not only shortens the computation time, but it may also improve the quality of the origin localization under certain circumstances. In the following paragraphs, we present a numerical estimation of the complexity of GMLA as well as its performance in terms of the quality of results in comparison to PTV.

We compare GMLA and PTV on four kinds of synthetic networks: Erdős-Rényi (ER), Barabási-Albert (BA), random regular graph (RRG) and scale-free network (SFN). We use well-known susceptible-infected as transmission model with the infection rate $\beta = 0.5$ (which corresponds to propagation ratio $\lambda = \mu/\sigma = \sqrt{2}$). The observers are distributed in random nodes with the density $\rho = 0.2$. For a high precision of GMLA, we set the number of the nearest observers as a sublinear function of the network size $K_0 = 0.5\sqrt{N}$ (see Fig. 2.5b). For comparative purposes, we introduce also a baseline method. The baseline method assumes that the true source is always the *observer one* (with smallest delay $t_k$). We use the average precision as a quality measure. Precision for a single test (realization) is defined as the ratio between the number of correctly located sources (i.e. true positives, which here equals either zero or one) and the number of sources found by the method (i.e. true positives plus false positives, which here is at least one). The tests are repeated multiple times for

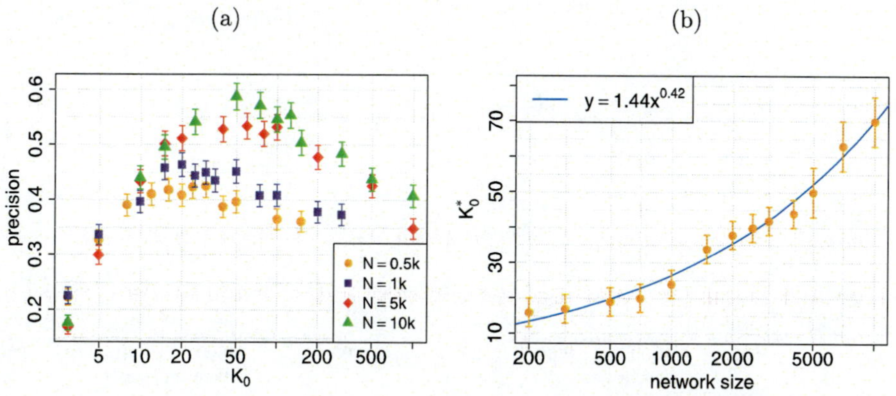


**Fig. 2.5** Performance of GMLA versus number of the nearest observers $K_0$ for various sizes of BA network ($m = 3$). **a** The maximum of precision is clearly visible. The position of maximum depends on the network size $N$. **b** The optimal number of the nearest observers $K_0^*$ for various sizes of BA network. The solid line is a nonlinear least squares model $y = bx^a$ (Gauss-Newton algorithm), where $a = 0.42 \pm 0.04$ and $b = 1.44 \pm 0.51$ (95% confidence interval)

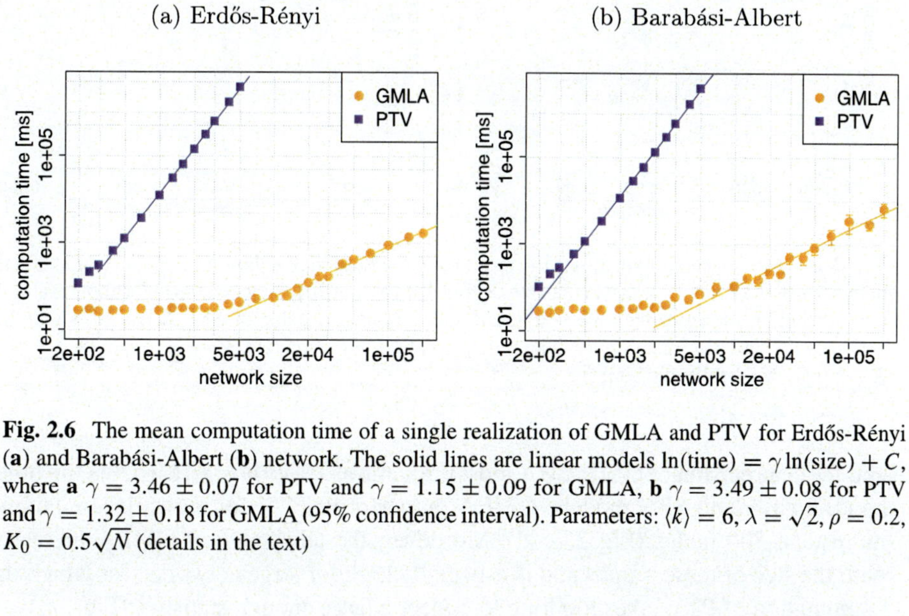


**Fig. 2.6** The mean computation time of a single realization of GMLA and PTV for Erdős-Rényi (**a**) and Barabási-Albert (**b**) network. The solid lines are linear models $\ln(\text{time}) = \gamma \ln(\text{size}) + C$, where **a** $\gamma = 3.46 \pm 0.07$ for PTV and $\gamma = 1.15 \pm 0.09$ for GMLA, **b** $\gamma = 3.49 \pm 0.08$ for PTV and $\gamma = 1.32 \pm 0.18$ for GMLA (95% confidence interval). Parameters: $\langle k \rangle = 6$, $\lambda = \sqrt{2}$, $\rho = 0.2$, $K_0 = 0.5\sqrt{N}$ (details in the text)

different sources and many graph realizations (for synthetic networks) and then the obtained values of precision are averaged.

The most important feature of GMLA is a striking reduction of the computation time. In Fig. 2.6, one can see that the empirical complexity decreases from $O(N^{3.46})$ to $O(N^{1.15})$ for ER graph and from $O(N^{3.49})$ to $O(N^{1.32})$ for BA network. Furthermore, an initial difference between GMLA and PTV computing times for the networks of size 200 is a factor 4.4 for ER graph and 3.6 for BA network.

As evident, the quality of the source localization depends on the network topology. Both algorithms perform better for ER and RRG than BA and SFN. In the case of ER

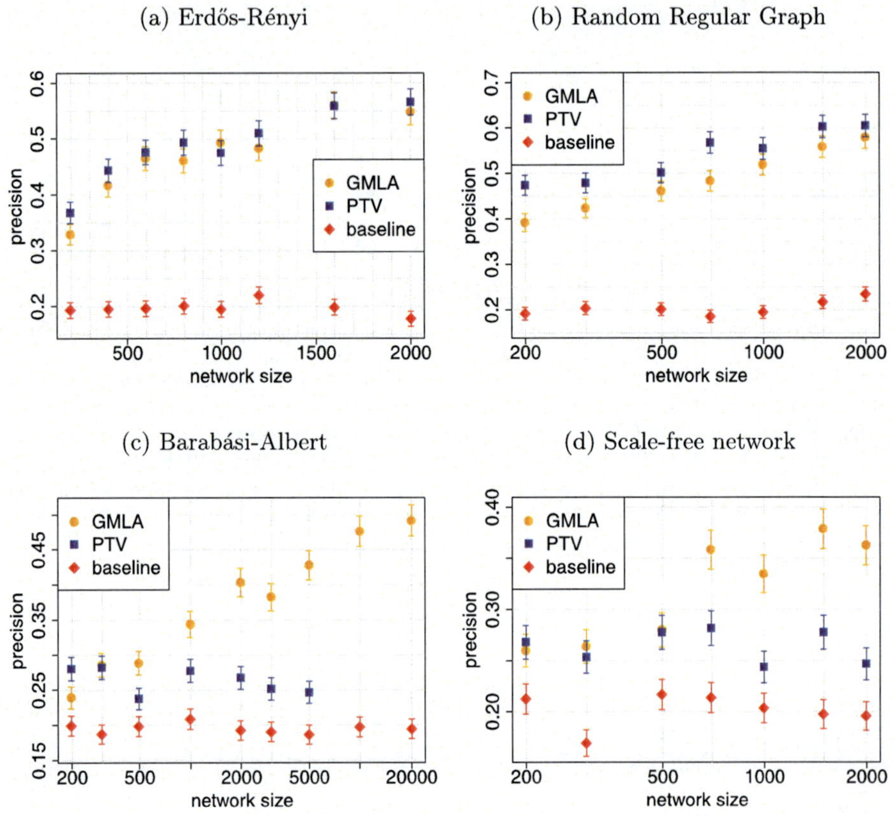


**Fig. 2.7** GMLA (orange) versus PTV (purple) and the baseline method (red) on several synthetic networks. **(a–b)** For networks with narrow degree distribution, precision of both algorithms is almost the same and it increases with the number of nodes. (**c–d**) Precision of GMLA increases with the network size and it is much higher than precision of PTV. Parameters: $\langle k \rangle = 6$, $\lambda = \sqrt{2}$, $\rho = 0.2$, $K_0 = 0.5\sqrt{N}$ (details in the text)

and RRG, precision of GMLA is almost the same as PTV (Fig. 2.7a–b). On the other hand, for BA and SFN models, GMLA outperforms PTV for networks containing more than 500 nodes (Fig. 2.7c–d). Moreover, the advantage of GMLA increases with the size of these graphs and is especially high for large networks, for which the computation of PTV takes too long to collect a large enough statistics (Fig. 2.6).

Figure 2.8 presents the precision of GMLA and PTV on a real peer-to-peer network. Gnutella is used for direct exchange of data via Internet between users and therefore can be used to spread the malware. The graph is downloaded from SNAP Datasets [30–32] and includes $N = 6299$ nodes ($\langle k \rangle = 6.6$). During tests we use simple SI model to simulate spreading. The performance of both methods is very similar for the density of the observers below 10%, but for higher densities GMLA provides higher precision of the source localization. Nevertheless, the main differ-

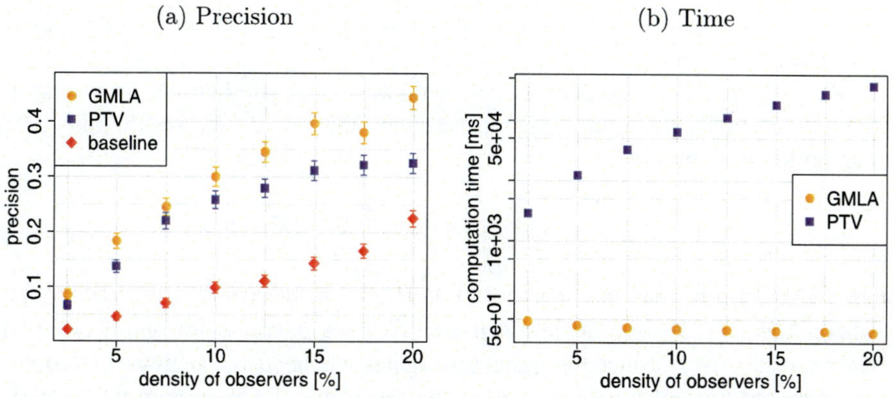


**Fig. 2.8** GMLA (orange) versus PTV (purple) and the baseline method (red) on Gnutella network. **a** GMLA achieves higher precision than PTV and the baseline method. **b** The mean computation time of a single realization of GMLA is substantially shorter that PTV and, in contrast to PTV, is independent of the density of observers. Each point in the figures is the result of 1000 realizations. Parameters: $\lambda = \sqrt{2}$, $K_0 = 30$ (details in the text)

ence between these algorithms lies in the computation time (Fig. 2.8b), which differs initially by a factor 61.5 and increases with the density of observers.

In summary, in this section, we proposed a novel approach for fast and accurate localization of spread source with incomplete observations which is capable to process timely large networks consisting of tens of thousands of nodes. Our algorithm is always much faster and provides higher quality of localization results than Pinto-Thiran-Vetterli algorithm for scale-free networks. The key to this success is restriction to the most important observers, while ignoring excessive and noisy information from far observers, as well as use of likelihood gradient for selection of potential origins of spread.

In the next section, we shall show how to improve the quality of detection results for networks that do not possess a tree topology and contain loops.

## Independent Paths

So far we have shown how to compensate the complexity of the PTV approach using a gradient like evaluation of node scores and limiting the number of observers considered in the evaluation. In this section, we will explain why obtaining expected traversal times exactly is a fairly complicated task (and therefore Pinto et al. focused on loop-free tree graphs) and how one can take into account more than just the shortest path between two nodes, i.e. multiple propagation paths. Finally, we will settle on using only the shortest paths, albeit *all of them* instead of simply one per pair of nodes as in PTV, and introduce two new ways—EPP and EPL—of computing the elements of (1).

We can represent each path $X_i$ between any given two nodes as a traversal time that is a normally distributed random variable $X_1, X_2, \ldots, X_n$ with a known joint probability distribution:

$$f(\boldsymbol{X}) = \frac{\exp\left(-\frac{1}{2}(\boldsymbol{X}-\boldsymbol{\mu})^T \boldsymbol{\Sigma}^{-1}(\boldsymbol{X}-\boldsymbol{\mu})\right)}{\sqrt{T^n|\boldsymbol{\Sigma}|}}, \tag{5}$$

where $\boldsymbol{X} = [X_1, X_2, \ldots, X_n]$, and $\boldsymbol{\mu}$ and $\boldsymbol{\Sigma}$ are a mean vector and a covariance matrix, respectively. Since the signal propagates via the *fastest* path we can compute the expected time of traversal as the expected value of a minimum distribution—$\mathbf{E}[X_{min}] = \mathbf{E}[\min(\boldsymbol{X})] = \mathbf{E}[\min(X_1, X_2, \ldots, X_n)]$ (Fig. 2.9).

We can write the probability density function (PDF) as

$$\phi(X_{min}) = -\frac{d}{dX_{min}} P(X_1, X_2, \ldots, X_n > X_{min}) = \tag{6}$$

$$= -\frac{d}{dX_{min}} \int\limits_{X_{min}}^{\infty} \cdots \int\limits_{X_{min}}^{\infty} f(\boldsymbol{X}) dX_1 dX_2 \ldots dX_n$$

and so the expected value $X_{min}$ is of course

$$\mathbf{E}[X_{min}] = \int\limits_{-\infty}^{\infty} x\phi(x)dx. \tag{7}$$

However, such an integral is not a trivial one to be fully computed analytically. There are two natural simplifications that can be used: ignoring correlations and limiting number of paths. Firstly, let us assume that there are no correlations among $X_i$, i.e. they are independently, identically distributed random variables (I.I.D.R.V). In such case, we can write the cumulative distribution function (CDF) of $X_{min}$ as follows:

$$\Phi_{X_{min}}(x) = P(X_{min} \leq x) = 1 - P(\min(X_1, X_2, \ldots, X_n) > x). \tag{8}$$

We also note an obvious fact that

$$\min(X_1, X_2, \ldots, X_n) > x \iff \forall_{i=1,2,\ldots,n} X_i > x, \tag{9}$$

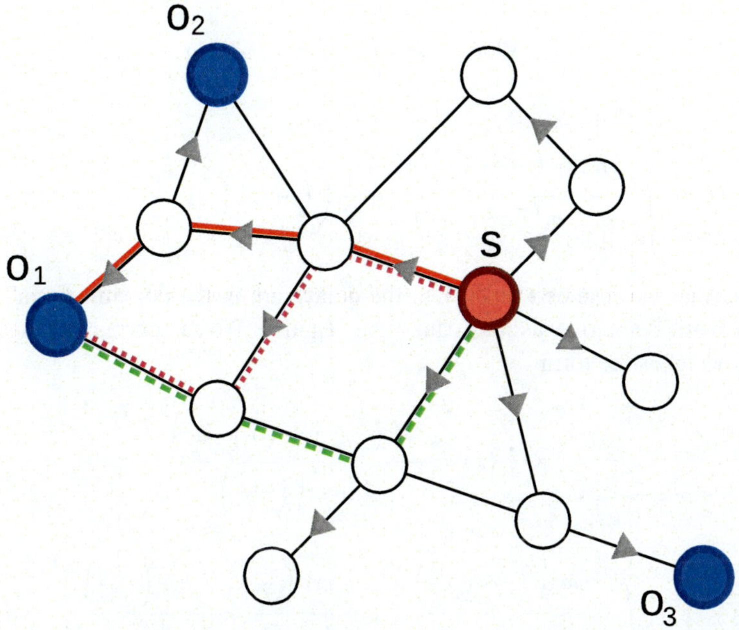


**Fig. 2.9** A schematic illustration of the problem. The source of the propagation is represented by the red node $S$ while observers by blue nodes $O_i$. There are three topologically shortest paths between $O_1$ and the $S$. To complicate the matters further these paths are correlated—some edges in the graph are shared. Additionally, the random nature of delays implies that a topologically longer route can still be the fastest what is shown with the observer $O_2$

which allows us to compute the CDF as

$$\Phi_{X_{min}}(x) = 1 - P(X_1 > x)P(X_2 > x)\dots P(X_n > x) = 1 - P(X_1 > x)^n = \\ = 1 - [1 - P(X_1 \leq x)]^n = 1 - \left[1 - \Phi_{X_1}(x)\right]^n, \tag{10}$$

where $\Phi_{X_1}$ has a known form:

$$\Phi_{X_1}(x) = \frac{1}{2}\left(1 + \text{erf}\left(\frac{x-\mu}{\sqrt{2}\sigma}\right)\right). \tag{11}$$

From which we can obtain the PDF of $X_{min}$:

$$\phi_{X_{min}}(x) = \frac{d\Phi_{X_{min}}(x)}{dx}, \tag{12}$$

which allows us to represent the expected value in an integral form:

$$\mathbf{E}[X_{min}] = \int_{-\infty}^{\infty} x\phi_{X_{min}}(x)dx = \tag{13}$$

$$= \int_{-\infty}^{\infty} \frac{xn\exp\left(-\frac{(x-\mu)^2}{2\sigma^2}\right)}{\sqrt{T}\sigma}\left[1-\frac{1}{2}\left(1+\text{erf}\left(\frac{x-\mu}{\sqrt{2}\sigma}\right)\right)\right]^{n-1} dx.$$

Note that we assumed so far that the delays are in the domain of real numbers. In real-world scenarios, a delay cannot be negative. We can correct for that and then the above takes the form:

$$\Phi_{X_1}(x) = \frac{\text{erf}\left(\frac{x-\mu}{\sqrt{2}\sigma}\right)+\text{erf}\left(\frac{\mu}{\sqrt{2}\sigma}\right)}{1+\text{erf}\left(\frac{\mu}{\sqrt{2}\sigma}\right)} \tag{14}$$

$$\mathbf{E}[X_{min}] = \int_{-\infty}^{\infty} \frac{xn\exp\left(-\frac{(x-\mu)^2}{2\sigma^2}\right)}{\sqrt{T}\sigma}\left[1-\frac{\text{erf}\left(\frac{x-\mu}{\sqrt{2}\sigma}\right)+\text{erf}\left(\frac{\mu}{\sqrt{2}\sigma}\right)}{1+\text{erf}\left(\frac{\mu}{\sqrt{2}\sigma}\right)}\right]^{n-1} dx. \tag{15}$$

It is also note worthy that this correction is not necessary when the *propagation ratio*—$\mu/\sigma$—is sufficiently large (due to the nature of *error function*).

In Fig. 2.10, we show the results of the above-described approach in contrast with other approaches, including the PTV tree approximation. These results show certain promise, however, as one increases the propagation ratio the assumption of uncorrelated paths is less justified. This makes this solution only viable in certain scenarios and there also exist another caveat of computational costs with numerical integration that often cannot be ignored.

## Multiple Correlated Paths

Now we will move onto the second approach—limiting number of paths taken into account. As indicated above, finding the general expression for the expected value of minimum distribution of correlated $n$ random variables is far from trivial [33]. Even with the I.I.D approximation the results force us to use numerical computation. There is, however, a known analytical formula in the case of $n = 2$ derived in [34].

Let $Y = min(X_1, X_2)$ where $X_1, X_2$ are known normal distributions with $E[X_i] = \mu_i$, $E[X_i^2] - E[X_i]^2 = \sigma_i$, and $\Phi, \phi$ representing the CDF and PDF of the standard normal distribution. Under this notation authors in [34] give

$$E[Y] = \mu_1\Phi\left(\frac{\mu_2-\mu_1}{\theta}\right)+\mu_2\Phi\left(\frac{\mu_1-\mu_2}{\theta}\right)-\theta\phi\left(\frac{\mu_2-\mu_1}{\theta}\right) \tag{16}$$

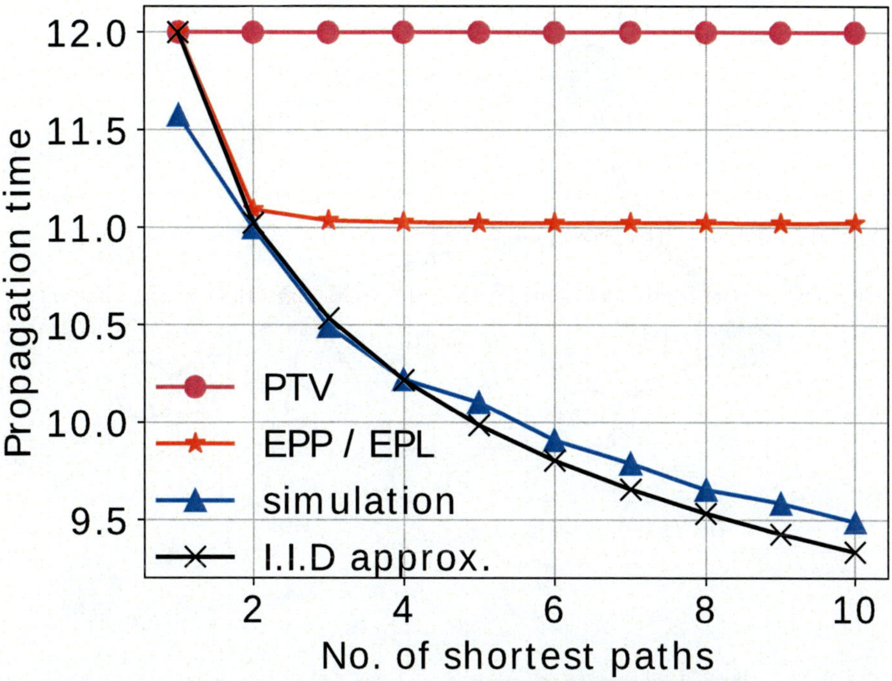


**Fig. 2.10** Signal's propagation time between two nodes in a network as a function of number of shortest paths connecting those nodes. Simulations were conducted on Barabàsi-Albert graph with 100 nodes, with propagation ratio $\frac{\mu}{\sigma} = 4$, the shortest path length was chosen to be three edges. Line labelled *simulation* is the real, i.e. experimental time whereas *PTV* is time assumed in the tree approximation. *EPP/EPL* is result of taking into account up to two shortest paths see (16). *I.I.D* is what one gets when taking all shortest paths into consideration with the assumption that they are uncorrelated

$$E[Y^2] = (\sigma_1^2 + \mu_1^2)\Phi\left(\frac{\mu_2 - \mu_1}{\theta}\right) + (\sigma_2^2 + \mu_2^2)\Phi\left(\frac{\mu_1 - \mu_2}{\theta}\right) - (\mu_1 + \mu_2)\theta\phi\left(\frac{\mu_2 - \mu_1}{\theta}\right) \quad (17)$$

$$\theta = \sqrt{\sigma_1^2 + \sigma_2^2 - 2\rho\sigma_1\sigma_2}. \quad (18)$$

With $\rho$ representing the correlation of $X_1$, $X_2$ that we shall define as the number of edges that the paths $R_1$, $R_2$ share with one another, normalized by their length $L$:

$$\rho = \frac{R_1 \cap R_2}{L}. \quad (19)$$

What we propose here is then to compute the expected times vector $\boldsymbol{\mu}$ using (16) if there is more than a single shortest path otherwise, of course, as in PTV. In case there are more than two shortest paths, apply (16) for two *least correlated* ones. In Fig. 2.10, we show the results of this proposition. It is quite apparent that this

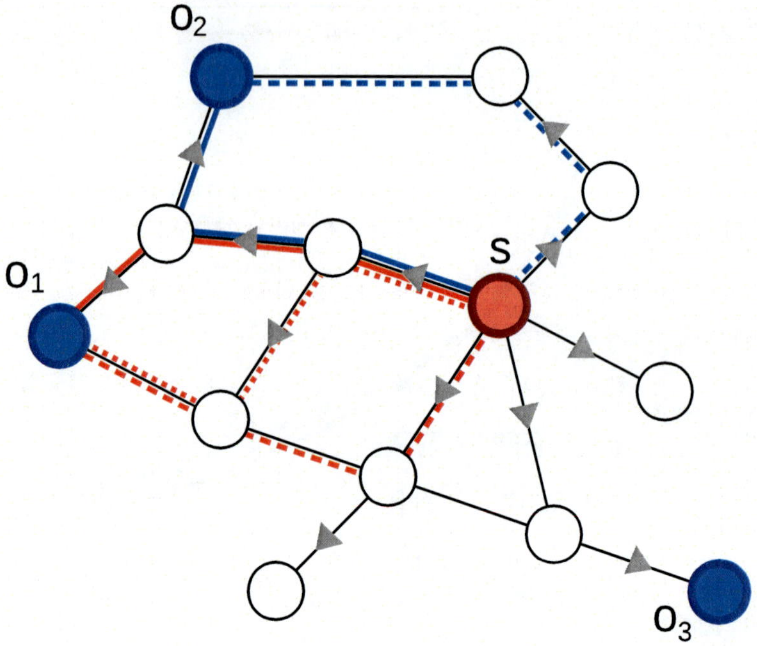


**Fig. 2.11** A schematic illustration of correlations of paths connecting the source $S$ and observers, $O_1$ (red paths) and $O_2$ (blue paths). Approach EPP assumes all paths are equally probable and computes the expected overlap between all pairs of red-blue paths (out of six pairs, one has overlap 1, one has overlap 2). On the other hand, approach EPL assumes all links are equally probable and computes the overlap between the union of all red and the union of all blue paths (2 links), normalized by the union of all paths (11 links)

deviates from reality as the number of shortest of paths grows; however, a significant improvement over the tree approximation is visible nonetheless.

## *Equiprobable Paths (EPP)*

We very deliberately reject the tree approximation used in PTV and therefore a new way of computing the covariance matrix is needed. We find it to be quite natural to generalize the formula used in PTV by the computation of the expected value of the covariance between all shortest paths from node $o_{i+1}$ to reference observer $o_1$ and all shortest paths between $o_{j+1}$ and $o_1$. The diagonal elements, on the other hand, instead of $\sigma^2 * L$ we put the variance of a minimum distribution if $n > 1$ (Fig. 2.11). Denote $\heartsuit$ as a set of paths each represented by a random variable:

$$\heartsuit_{i+1} := R_1(o_1, o_{i+1}), R_2(o_1, o_{i+1}), \ldots, R_n(o_1, o_{i+1}), \tag{20}$$

with $R_i(x, y)$ signifying an $i$th path between vertices $x$ and $y$, then

$$\Sigma_{i,j} = \begin{cases} \sigma^2 \sum_{k=1}^{n} \sum_{l=1}^{m} p_{k,l} \cdot |R_k(o_1, o_{i+1}) \cap R_l(o_1, o_{j+1})| & i \neq j \\ \min_{\sigma^2}(\heartsuit_{i+1}) = E[Y^2] - E[Y]^2 & i = j \end{cases}. \tag{21}$$

Here $p_{k,l}$ is a probability that $k$th path from the set of paths connecting nodes $o_1, o_{i+1}$ and $l$th path from the set connecting the other observer with reference observer (i.e. $o_1, o_{j+1}$) will be the fastest routes, see Fig. 2.11 for details. As it happens these probabilities are as difficult to compute as the expected traversal times themselves and we are forced to assume that all paths are equally probable (which is true only when there are no intersections between them). This assumption allows us to write

$$\Sigma_{i,j} = \begin{cases} \frac{\sigma^2}{n \cdot m} \sum_{k=1}^{n} \sum_{l=1}^{m} |R_k(o_1, o_{i+1}) \cap R_l(o_1, o_{j+1})| & i \neq j \\ \min_{\sigma^2}(\heartsuit_{i+1}) & i = j \end{cases} \tag{22}$$

## *Equiprobable Links (EPL)*

Our second proposition of computing the covariance matrix is to take a product of the Jaccard Index [35] (Fig. 2.11) and the geometric mean of variances of sets of paths, namely,

$$\Sigma_{i,j} = \frac{|\{e_i\} \cap \{e_j\}|}{|\{e_i\} \cup \{e_j\}|} \cdot \sqrt{\min_{\sigma^2}(\heartsuit_{i+1}) \cdot \min_{\sigma^2}(\heartsuit_{j+1})}, \tag{23}$$

where $\{e_i\}$ is a set of edges of all shortest paths connecting node $i$ with the reference observer. The intuition behind this approach is to say that using *intersection over union* of sets of edges we treat each edge as equally probable (in contrast to treating each *path* equally probable like in the previous approach). The choice of geometric mean here is fairly arbitrary and we found arithmetic mean to produce similar results.

## *EPP/EPL Results*

Our method presented above has been tested on both synthetic graphs and real networks. We used two synthetic graphs: Barabáisi-Albert (BA) and Erdős-Rényi (ER) with size $N = 100$ and average node degree $\langle k \rangle = 8$ in each case. As a measure of success we use precision like in the previous sections. Results are shown in Fig. 2.12. There is visible a vast improvement in precision of the method using our proposed adjustments with the difference between EPP and EPL being barely noticeable.

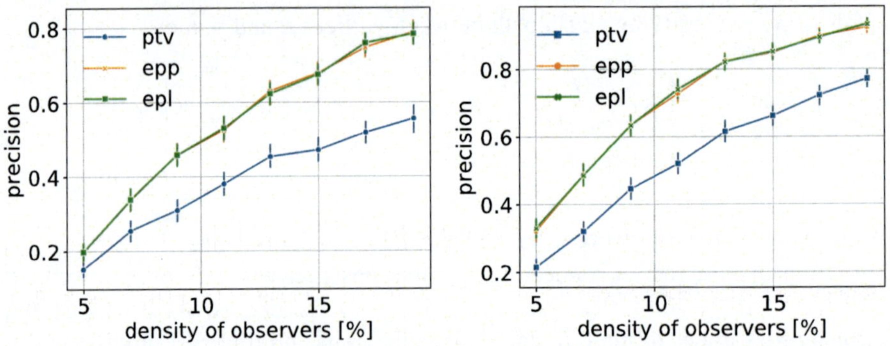


(a) Barabasi-Albert (b) Erdos-Renyi

**Fig. 2.12** Improved precision of source detection measured as a function of observers' density for original PTV method and our modifications. Simulations were conducted on **a** Barabási-Albert (BA) and **b** Erdős-Rényi (ER) graphs. Both with mean node degree $\langle k \rangle = 8$ and network size $N = 100$. We have conducted 1000 simulations for each point on those plots. Bars represent 95% confidence intervals

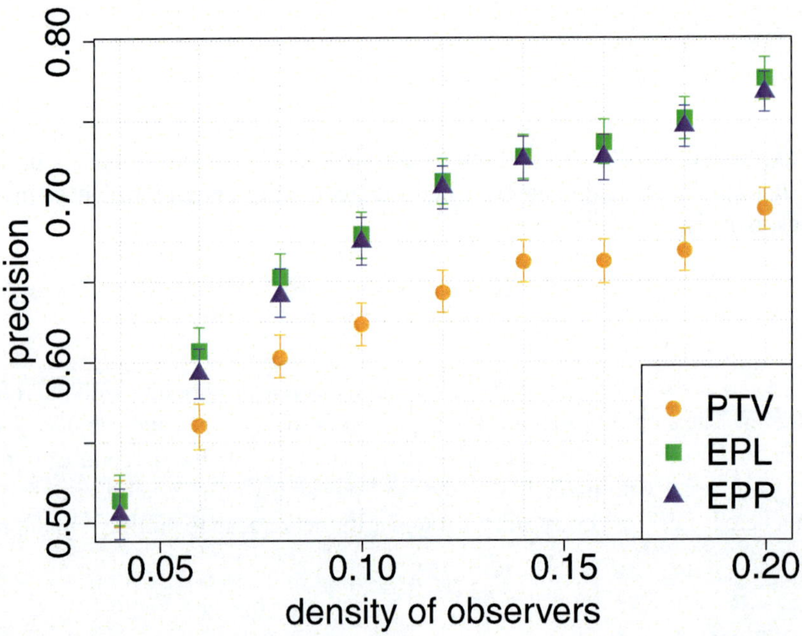


**Fig. 2.13** Precision gain on the University of Rovira i Virgili network. The network has $N = 1133$ nodes, mean degree $\langle k \rangle = 9.6$ and clustering coefficient $C = 16.6\%$. The propagation times are distributed normally with propagation ratio $\mu/\sigma = 4$. The results are averaged over 1000 realizations. Bars represent 95% confidence intervals

We have also conducted a study on a real network from the University of Rovira i Virgili. The network has $N = 1133$ nodes, mean degree $\langle k \rangle = 9.6$ and clustering coefficient $C = 16.6\%$ [36]. It is the email communication network at the University Rovira i Virgili in Tarragona in the south of Catalonia in Spain. Nodes are users and each edge represents that at least one email was sent. The direction of emails or the number of emails are not known [37]. The results are shown in Fig. 2.13 with each point being a result of 1000 simulations. One can clearly see a confirmation of the previous findings on synthetic networks—both our modifications show substantial increase in precision with EPL being ever so slightly better than EPP. The fact that this result holds for real-world network strongly shows the merit of our methods for real-world applications.

## Conclusions

We introduce a new algorithm (GMLA) for the spread source localization in the well-known Pinto-Thiran-Vetterli (PTV) limited observers formulation. The main drawback of PTV is its time complexity. For large networks with many observers, the complexity of PTV is defined by the complexity of matrix operations, which is $O(K^3)$ per node in the worst case (where $K$ denotes the number of observers). We avoid this drawback in our algorithm by reducing the number of observers used to determine the score (the likelihood of a node being the source) and by limiting the number of suspected nodes. The latter is performed by the selection procedure which starts from the neighbours of the first observer and follows the gradient of the score. As a result of this selection, we get a limited number of the suspected nodes $N_0 = |V_s| \sim \log N$ in contrast to PTV where each node is checked ($V_s = V$). Thanks to this approach, the complexity of gradient maximum likelihood algorithm (GMLA) is $O(\log(N)N^2)$ in the worst case and as far as we know this is the fastest algorithm for the spread source detection in generic networks with incomplete observations. The phrase "less is more" once again turned out to be truth here.

We also address the issue of a tree-like propagation assumption that is often not true. We have shown that the approach of breadth-first search ($BFS$) trees presented by Pinto et al. overestimates propagation time of a signal. We presented an analysis of why that is, namely, there can be more than one shortest paths and also other paths may also happen to be faster, and those effects are non-negligible. We provide alternatives to $BFS$ that take multiple paths into account and while they require more computational time, they significantly improve precision of the source detection. This improvement is the most prominent for artificial networks; however, when tested on a real network there was also a significant increase in precision with both EPL and EPP over the PTV method. To our knowledge, our approaches (EPL/EPP) provide highest precision of location among current methods.

The two approaches presented here address on of the most important challenges in context of maximum likelihood source estimation. We show how to modify the PTV method to some specific needs in various scenarios. If the most important consideration is the time of computation—one can use the GMLA variant. Should precision be of highest importance one can use EPL or EPP. Which of the two to use in the latter case also depends on whether we have additional information we could use (e.g. the path probabilities in EPP). While our methods do not completely solve the problem of locating the source of spread in complex networks, we do hope they provide an appropriate toolbox for the reader that will prove useful in the real world.

**Acknowledgements** The work was partially supported as RENOIR Project by the European Union Horizon 2020 research and innovation programme under the Marie Skłodowska-Curie grant agreement no. 691152 and by Ministry of Science and Higher Education (Poland) grant nos. 34/H2020/2016, 329025/PnH /2016 National Science Centre, Poland grant no. 2015/19/B/ST6/02612 and by POB Research Centre Cybersecurity and Data Science of Warsaw University of Technology within the Excellence Initiative Program - Research University (IDUB). J.A.H. was partially supported by the Russian Scientific Foundation, Agreement #17-71-30029 with co-financing of Bank Saint Petersburg. B.K.S. was partially supported by the Army Research Laboratory under Cooperative Agreement Number W911NF-09-2-0053 (the ARL Network Science CTA), by the Army Research Office grant W911NF-16-1-0524 and by the Office of Naval Research grant N00014-15-1-2640. R.P. was partially supported by the National Science Centre, Poland, agreement no. 2019/32/T/ST6/00173 and by PLGrid Infrastructure.

## References

1. A.L. Barabási, *Linked: How Everything Is Connected to Everything Else and What It Means for Business, Science, and Everyday Life* (Plume, New York, 2003)
2. M.E.J. Newman, The structure and function of complex networks. SIAM Rev. **45**(2), 167–256 (2003)
3. D. Helbing, S. Balietti, From social data mining to forecasting socio-economic crises. Eur. Phys. J. Spec. Top. **195**(1), 3 (2011)
4. F. Giannotti, D. Pedreschi, A. Pentland, P. Lukowicz, D. Kossmann, J. Crowley, D. Helbing, A planetary nervous system for social mining and collective awareness. Eur. Phys. J. Spec. Top. **214**(1), 49–75 (2012)
5. R. Pastor-Satorras, A. Vespignani, Epidemic spreading in scale-free networks. Phys. Rev. Lett. **86**, 3200–3203 (2001)
6. I. Moya, M. Chica, J.L. Saez-Lozano, O. Cordon, An agent-based model for understanding the influence of the 11-M terrorist attacks on the 2004 Spanish elections. Knowl.-Based Syst. **123**, 200–216 (2017)
7. M. Sun, H. Zhang, H. Kang, G. Zhu, X. Fu, Epidemic spreading on adaptively weighted scale-free networks. J. Math. Biol. **74**(5), 1263–1298 (2017)
8. F. Fu, N.A. Christakis, J.H. Fowler, Dueling biological and social contagions. Sci. Rep. 7 (2017)
9. G. Strauss, A. Shell, R. Yu, B. Acohido, SEC, FBI probe fake tweet that rocked stocks. USA Today (2013)
10. H. Alcott, M. Gentzkow, Social media and fake news in the 2016 election. J. Econ. Perspect. **31**(2), 211–236 (2017)
11. D. Shah, T. Zaman, Rumors in a network: who's the culprit? IEEE Trans. Inf. Theory **57**(8), 5163–5181 (2011)
12. P.C. Pinto, P. Thiran, M. Vetterli, Locating the source of diffusion in large-scale networks. Phys. Rev. Lett. **109**(6), 1–5 (2012)
13. B.A. Prakash, J. Vrekeen, C. Faloutsos, Spotting culprits in epidemics: how many and which ones? in *Proceedings - IEEE International Conference on Data Mining, ICDM* (2012), pp. 11–20

14. A.Y. Lokhov, M. Mézard, H. Ohta, L. Zdeborová, Inferring the origin of an epidemic with a dynamic message-passing algorithm. Phys. Rev. E - Stat. Nonlinear Soft Matter Phys. **90**(1), 1–9 (2014)
15. K. Zhu, L. Ying, Information source detection in the SIR model: a sample-path-based approach. IEEE/ACM Trans. Netw. **24**(1), 408–421 (2016)
16. N. Karamchandani, M. Franceschetti, Rumor source detection under probabilistic sampling, in *IEEE International Symposium on Information Theory - Proceedings* (2013), pp. 2184–2188
17. W. Luo, W.P. Tay, M. Leng, How to identify an infection source with limited observations. IEEE J. Sel. Top. Signal Process. **8**(4), 586–597 (2014)
18. D. Brockmann, D. Helbing, The hidden geometry of complex, network-driven contagion phenomena. Science **342**(6164), 1337–1342 (2013)
19. N. Antulov-Fantulin, A. Lančić, T. Šmuc, H. Štefančić, M. Šikić, Identification of patient zero in static and temporal networks: robustness and limitations. Phys. Rev. Lett. **114**(24), 1–5 (2015)
20. Z. Shen, S. Cao, W.X. Wang, Z. Di, H.E. Stanley, Locating the source of diffusion in complex networks by time-reversal backward spreading. Phys. Rev. E - Stat. Nonlinear Soft Matter Phys. **93**(3), 1–9 (2016)
21. A. Braunstein, A. Ingrosso, Inference of causality in epidemics on temporal contact networks. Sci. Rep. **6**, 27538 (2016)
22. J. Jiang, S. Wen, S. Yu, Y. Xiang, W. Zhou, Rumor source identification in social networks with time-varying topology. IEEE Trans. Dependable Secur. Comput. **5971**(c), 1–1 (2016)
23. L. Fu, Z.S. Shen, W.X. Wang, Y. Fan, Z.R. Di, Multi-source localization on complex networks with limited observers. EPL **113**(1) (2016)
24. V. Fioriti, M. Chinnici, J. Palomo, Predicting the sources of an outbreak with a spectral technique. Appl. Math. Sci. **8**(133–136), 6775–6782 (2014)
25. J. Jiang, S. Wen, S. Yu, Y. Xiang, W. Zhou, Identifying propagation sources in networks: state-of-the-art and comparative studies. IEEE Commun. Surv. Tutor. **X**(X), 1–17 (2014)
26. R. Paluch, X. Lu , K. Suchecki, B.K. Szymański, J.A. Hołyst, Fast and accurate detection of spread source in large complex networks. Sci. Rep. **8**(1), 2508 (2018)
27. Ł.G. Gajewski, K. Suchecki, J.A. Hołyst, Multiple propagation paths enhance locating the source of diffusion in complex networks. Phys. A: Stat. Mech. Its Appl. **519**, 34–41 (2019)
28. B. Spinelli, L.E. Celis, P. Thiran, Observer placement for source localization: the effect of budgets and transmission variance, in *54th Annual Allerton Conference on Communication, Control, and Computing (Allerton)* (2016), pp. 743–751
29. R. Albert, A.L. Barabási, Statistical mechanics of complex networks. Rev. Mod. Phys. **74**(1), 47–97 (2002)
30. J. Leskovec, A. Krevl, Gnutella peer-to-peer network: snapshot from August 8, 2002, http://snap.stanford.edu/data/p2p-Gnutella08.html. Accessed 30 Nov 2017
31. R. Matei, A. Iamnitchi, I. Foster, Mapping the Gnutella network. IEEE Internet Comput. **6**(1), 50–57 (2002)
32. J. Leskovec, J. Kleinberg, C. Faloutsos, Graph evolution: densification and shrinking diameters. ACM Trans. Knowl. Discov. Data **1**(1) (2007)
33. R.B. Arellano-Valle, M.G. Genton, On the exact distribution of the maximum of absolutely continuous dependent random variables. Stat. Probab. Lett. **78**(1), 27–35 (2008)
34. S. Nadarajah, S. Kotz, Exact distribution of the max/min of two gaussian random variables. IEEE Trans. Very Large Scale Integr. (VLSI) Syst. **16**, 210–212 (2008)
35. P. Jaccard, Le coefficient generique et le coefficient de communaute dans la flore marocaine. Mém. Soc. Vaud. Sci. Nat. **2**, 385–403 (1926)
36. R. Guimerà, L. Danon, A. Díaz-Guilera, F. Giralt, A. Arenas, Self-similar community structure in a network of human interactions. Phys. Rev. E **68**(6), 065103 (2003)
37. U. Rovira i Virgili network dataset – KONECT (2017)

# Chapter 3
# Power-Law Cross-Correlations: Issues, Solutions and Future Challenges

**Ladislav Kristoufek**

**Abstract** Analysis of long-range dependence in financial time series was one of the initial steps of econophysics into the domain of mainstream finance and financial economics in the 1990s. Since then, many different financial series have been analysed using the methods standardly used outside of finance to deliver some important stylized facts of the financial markets. In the late 2000s, these methods have started being generalized to bivariate settings so that the relationship between two series could be examined in more detail. It was then only a single step from bivariate long-range dependence towards scale-specific correlations and regressions as well as power-law coherency as a unique relationship between power-law correlated series. Such rapid development in the field has brought some issues and challenges that need further discussion and attention. We shortly review the development and historical steps from long-range dependence to bivariate generalizations and connected methods, focus on its technical aspects and discuss problematic parts and challenges for future directions in this specific subfield of econophysics.

## Introduction

Analysis of long-range dependence properties of financial time series was at the very beginning of the econophysics field in the early 1990s [2, 28, 34] following the early works of the Mandelbrot research group [25–27]. At the time, most of the financial works were based on the assumption that, in addition to other simplifying restrictions, the autocorrelation function of the series vanishes exponentially, i.e. very quickly. Lagged observations of the series thus played only a marginal role and only after few time steps, the effect was assumed to be gone completely. Such assumption has some convenient mathematical properties in a parallel logic to assuming the Gaussian distribution. However, the noted works, among others, have argued that

---

L. Kristoufek (✉)
Institute of Information Theory and Automation, Czech Academy of Sciences, Pod Vodarenskou vezi 4, Prague, Czech Republic
e-mail: kristouf@utia.cas.cz



D. Grech and J. Miśkiewicz (eds.), *Simplicity of Complexity in Economic and Social Systems*, Springer Proceedings in Complexity,
https://doi.org/10.1007/978-3-030-56160-4_3

some financial time series show that observations at even very high lags can have an effect on current movements of the financial series. This gave rise to the so-called Hurst effect with respect to [9]—and his work in hydrology—which has since been referred to by various names, mostly persistence, long-range dependence and long-term correlations (and sometimes long-term memory).

Long-range dependence of time series is characterized by a slowly decaying autocorrelation function, contrary to the quickly vanishing exponentially decreasing autocorrelation function standardly seen in autoregressive (integrated) moving-average processes (ARMA/ARIMA) [4] and (generalized) autoregressive conditional heteroskedasticity models (ARCH/GARCH) [3, 6] that are standard in financial economics and financial econometrics. In the econophysics literature, the slowly decaying autocorrelation function is usually represented by a hyperbolical decay. Such specification has some intriguing properties [2, 34] which allowed for introduction of many estimators of long-range dependence parameters, but most importantly to the detrended fluctuation analysis (DFA) [30, 31], which quickly became the most popular method of studying long-range dependence properties in the time domain. Its simplicity and intuitive appeal made it an ideal candidate for various specifications, adjustments and generalizations—most notably the multifractal detrended fluctuation analysis (MF-DFA) [10] and detrended cross-correlation analysis (DCCA) [33]. The former method generalizes the original one by studying multifractal properties rather than (mono/uni) fractal ones and the latter studies the long-range dependence properties between two series, i.e. cross-correlations rather than serial (auto-)correlations.

In this work, we study and review the methodological steps that needed to be taken when coming from long-range correlations towards long-range cross-correlations. Importantly, we focus on problematic parts of the latter and cover two approaches on how to treat them. Specifically, we argue (and review the relevant literature that shows so) that long-range (power-law) cross-correlations are only an in-between step and by themselves they tell very little. The two approaches, which utilize the power-law cross-correlations as the mentioned in-between step, are the scale-specific correlations and regressions, and the power-law coherency. Eventually, we show that these two are inherently related. In our discussion, we outline possible future challenges in this branch of interdisciplinary research.

## From Long-Range Dependence to Power-Law Cross-Correlations

Persistent series can be characterized through its dynamic properties in both time and frequency domains. In the former, the autocorrelation function is standardly represented by an asymptotic hyperbolic decay, specifically $\rho(k) \propto k^{2H-2}$ for $k \rightarrow +\infty$ where $\rho(k)$ is the autocorrelation function at time lag $k$ and $H$ is the Hurst exponent [34]. In the latter, the persistence translates into a power-law divergence of the spectrum at origin, specifically $f(\omega) \propto \omega^{1-2H}$ for $\omega \rightarrow 0+$ where $f(\omega)$ is the spectrum and $\omega$ is the frequency [2]. The critical parameter here is the Hurst exponent $H$. For the stationary series, the exponent ranges between 0 and 1 and is well separated by 0.5 which marks a process with no persistence. Processes with $H > 0.5$ are the persistent ones that have strong autocorrelation structure and remind of locally trending processes that still remain stationary. Antipersistent processes with $H < 0.5$ are characterized by frequent switching of signs of their changes but are usually of a marginal interest compared to the persistent processes that can be exploited in finance for profitable trading strategies [26]. Even though the frequency-domain approach has gained more traction in the financial econometrics field, it has been the time-domain estimators that became more popular in the interdisciplinary research. We follow this logic and focus on the time-domain implications of long-range dependence (even though most of it can be quite easily translated into the frequency-domain language).

The hyperbolic decay of the autocorrelation function has some interesting implications which are covered in various textbooks (we refer here to the "classics" of [2, 34]) but specifically its connection to the scaling of partial sums has crucial application. We define a partial sum of process $\{x_t\}_{t=1}^{T}$, where $T$ is the time series length, as $X_t = \sum_{i=1}^{t} x_i$. If $\{x_t\}_{t=1}^{T}$ is long-range correlated, then the variance of partial sums scales as $\text{Var}(X_t) \propto t^{2H}$ for $t \rightarrow +\infty$. It turns out that variance of an integrated process (the partial sum) is much less noisy than an autocorrelation function of the original process at high lags, which in turn makes the approach based on the partial sums and variance more appropriate for the Hurst exponent $H$ estimation.

The partial sums divergence is utilized in various estimators of the Hurst exponent, most notably in the detrended fluctuation analysis (DFA) [10, 30, 31]. DFA is based on several steps mainly focused on a further reduction of the noise in the estimation procedure as well as filtering out possible time trends. Specifically, one starts with a profile of the series (a cumulative sum of the de-meaned original series), which represents the cumulative sum in the previous paragraph. Such profile is split into intervals of length $s$ representing a scale. In each interval of the given length, a time trend is estimated (usually a linear trend but the procedure can utilize many different filtering procedures) and a mean squared error around the trend is found. This squared error is then averaged over all intervals of the given length to give a fluctuation function $F^2(s)$. The procedure is repeated for a range of scales and the Hurst exponent is estimated on the scaling rule $F^2(s) \propto s^{2H}$. DFA has become and remained the most popular of the time-domain Hurst exponent estimators even over its weaknesses as reported in various studies [1, 7, 11, 37–39] mainly due to its straightforward nature and implementation. Although it needs to be stressed that DFA is also the most tested and numerically examined of the time-domain Hurst exponent estimators.

It took more than a decade to come from DFA to a parallel examination of dependence between two series. And again, it was DFA in the centre. Podobnik and Stanley [33] introduced the detrended cross-correlation analysis (DCCA/DXA) that is built on a parallel idea—scaling of covariances between the partial sums. Even though the step from DFA to DCCA is intuitively clear and frankly trivial—instead of finding a mean squared error from the trend in each window of size $s$, we find a product of errors from the trend for two series—it took five more years to prove that, in fact, the covariance of partial sums of two series, out of which at least one is long-range correlated, scales as $\text{Cov}(X_t, Y_t) \propto t^{2H_{xy}}$ where $H_{xy}$ is the bivariate Hurst exponent [13]. And even though DCCA has quickly become popular and widely used in the empirical literature across disciplines, the theoretical paper of [13] was one of the first to show that going from univariate to bivariate perspective has some serious methodological caveats and making intuitive translations from the former to the latter without proper theoretical treatment can lead to crucial errors.

## The Issues with Power-Law Cross-Correlations

Most of the literature building on the DCCA procedure has been empirical and it has become quickly clear that the relationship between the bivariate Hurst exponent $H_{xy}$ and the Hurst exponents of the separate processes $H_x$ and $H_y$ might play a crucial role. From one side, most empirical studies reported that either $H_{xy} = \frac{1}{2}(H_x + H_y)$ or $H_{xy} > \frac{1}{2}(H_x + H_y)$ [8, 29, 40]. From the other, numerical and theoretical studies suggested that either $H_{xy} = \frac{1}{2}(H_x + H_y)$ or $H_{xy} < \frac{1}{2}(H_x + H_y)$ [36]. The clash was apparent and a more detailed theoretical treatment was clearly needed.

The primary issue of the literature (both theoretical and empirical) on the power-law cross-correlations was the non-existence of a process that would generate the power-law cross-correlated series and allow to control the $H_{xy}$ parameter. Even though [32] proposed the two-component ARFIMA process as a mixture of two power-law autocorrelated processes, it has not been numerically shown how to control the $H_{xy}$ parameter as a function of $H_x$, $H_y$ and the proposed weight $W$. The proposed process was verified by the DCCA estimation, which, unfortunately, is not a proper way of proving validity as DCCA itself has not been shown to have clear statistical properties. This circular proof is thus not valid.

Kristoufek [12] introduced the mixed-correlated ARFIMA process (MC-ARFIMA), which allowed for controlling the $H_{xy}$ parameter. MC-ARFIMA processes are defined as

$$x_t = \alpha \sum_{n=0}^{+\infty} a_n(d_1)\varepsilon_{1,t-n} + \beta \sum_{n=0}^{+\infty} a_n(d_2)\varepsilon_{2,t-n}$$
$$y_t = \gamma \sum_{n=0}^{+\infty} a_n(d_3)\varepsilon_{3,t-n} + \delta \sum_{n=0}^{+\infty} a_n(d_4)\varepsilon_{4,t-n},$$

where

$$a_n(d) = \frac{\Gamma(n+d)}{\Gamma(n+1)\Gamma(d)}$$

and error terms are characterized by

$$\begin{aligned}\langle \varepsilon_{i,t} \rangle &= 0 \text{ for } i = 1, 2, 3, 4 \\ \langle \varepsilon_{i,t}^2 \rangle &= \sigma_{\varepsilon_i}^2 \text{ for } i = 1, 2, 3, 4 \\ \langle \varepsilon_{i,t}\varepsilon_{j,t-n} \rangle &= 0 \text{ for } n \neq 0 \text{ and } i, j = 1, 2, 3, 4 \\ \langle \varepsilon_{i,t}\varepsilon_{j,t} \rangle &= \sigma_{ij} \text{ for } i, j = 1, 2, 3, 4 \text{ and } i \neq j.\end{aligned}$$

To put it in words, the two processes are each a linear combination of two power-law autocorrelated processes with possibly correlated error terms. The separate long-range dependence parameters $d_1, d_2, d_3, d_4$ are unrestricted. The $d$-notation is kept here mainly due to the use of the $\Gamma(\bullet)$ function and it standardly holds that $H = d + \frac{1}{2}$. Even though the paper discusses more possibilities, there are two important specifications. First, if we do not restrict the correlation between error terms in any way, the bivariate Hurst exponent will be an average of the separate Hurst exponents. And second, if the two processes with lower separate Hurst exponents in each have correlated error terms and the other error terms are uncorrelated, the bivariate Hurst exponent will be lower than the average of the two separate ones. There is no combination of parameters that would allow the bivariate Hurst exponent to be higher than the average of the separate ones. This can be quite easily seen from the asymptotic behaviour of the cross-correlation function between two MC-ARFIMA processes (and more details are given in the reference):

$$\begin{aligned}\rho_{xy}(n) = \cdots \approx \\ \frac{\alpha\gamma\sigma_{13}}{\sigma_x\sigma_y} \underbrace{\sum_{k=0}^{+\infty} a_k(d_1)a_{n+k}(d_3)}_{\approx \int_0^{+\infty} k^{d_1-1}(n+k)^{d_3-1}dk \propto n^{d_1+d_3-1}} + \frac{\alpha\delta\sigma_{14}}{\sigma_x\sigma_y} \underbrace{\sum_{k=0}^{+\infty} a_k(d_1)a_{n+k}(d_4)}_{\approx \int_0^{+\infty} k^{d_1-1}(n+k)^{d_4-1}dk \propto n^{d_1+d_4-1}} + \\ \frac{\beta\gamma\sigma_{23}}{\sigma_x\sigma_y} \underbrace{\sum_{k=0}^{+\infty} a_k(d_2)a_{n+k}(d_3)}_{\approx \int_0^{+\infty} k^{d_2-1}(n+k)^{d_3-1}dk \propto n^{d_2+d_3-1}} + \frac{\beta\delta\sigma_{24}}{\sigma_x\sigma_y} \underbrace{\sum_{k=0}^{+\infty} a_k(d_2)a_{n+k}(d_4)}_{\approx \int_0^{+\infty} k^{d_2-1}(n+k)^{d_4-1}dk \propto n^{d_2+d_4-1}}.\end{aligned}$$

The MC-ARFIMA introduction has had two main results. First, there was finally a data generator that could be used for simulation studies that also has well-defined statistical properties [17, 21]. And second, the possibility of having $H_{xy} > \frac{1}{2}(H_x + H_y)$ seemed to have vanished as the MC-ARFIMA processes are very generally defined and allow for very flexible manipulation. In other words, if it was not possible to find a specification that would lead to $H_{xy} > \frac{1}{2}(H_x + H_y)$ in this setting, it might be unattainable completely.

As a follow-up, [19] studies the issue of $H_{xy} > \frac{1}{2}(H_x + H_y)$ on a theoretical basis in more detail. As it turns out, the answer is almost trivial. The issue is solved through the squared spectrum coherency and its scaling close to the origin. The squared spectrum coherency is defined for two stationary series $\{x_t\}_{t=1}^T$ and $\{y_t\}_{t=1}^T$ with (cross-)spectra $f_{xy}(\omega)$, $f_x(\omega)$ and $f_y(\omega)$ at frequency $0 \leq \omega \leq \pi$ as

$$K_{xy}^2(\omega) = \frac{|f_{xy}(\omega)|^2}{f_x(\omega)f_y(\omega)}$$

for a given frequency $\omega$. Using the definition of the power-law cross-correlations in the frequency domain, we can rewrite the coherency as

$$K_{xy}^2(\omega) = \frac{|f_{xy}(\omega)|^2}{f_x(\omega)f_y(\omega)} \propto \frac{\omega^{2(1-2H_{xy})}}{\omega^{1-2H_x}\omega^{1-2H_y}} = \omega^{2(H_x+H_y-2H_{xy})}.$$

Now note that the squared coherency ranges between 0 and 1 everywhere (in fact even for non-stationary series with their pseudo-spectra). Therefore, it is so restricted for the long-range cross-correlations frequencies as well, i.e. $\omega \to 0+$. This gives us two feasible and one infeasible possibilities:

- $H_{xy} = \frac{1}{2}(H_x + H_y) \Rightarrow 2(H_x + H_y - 2H_{xy}) = 0 \Rightarrow \lim_{\omega\to 0+} K_{xy}^2(\omega) \propto \text{const.}$
- $H_{xy} < \frac{1}{2}(H_x + H_y) \Rightarrow 2(H_x + H_y - 2H_{xy}) > 0 \Rightarrow \lim_{\omega\to 0+} K_{xy}^2(\omega) = 0$
- $H_{xy} > \frac{1}{2}(H_x + H_y) \Rightarrow 2(H_x + H_y - 2H_{xy}) < 0 \Rightarrow \lim_{\omega\to 0+} K_{xy}^2(\omega) = +\infty \Rightarrow$

This implies that $H_{xy} > \frac{1}{2}(H_x + H_y)$ is impossible. Note that this holds for stationary as well as for non-stationary processes (and it can be easily shown for the DCCA fluctuations scaling as well). If the empirical literature reports otherwise, it is due to biased estimators. This bias might be due to various reasons. First, the standardly used estimators of the bivariate Hurst exponent $H_{xy}$ seem to be biased in general as well as due to short-term dependence bias [17] even though the latter should not be the case, at least theoretically [20]. Second, the estimators are strongly upward biased in presence of heavy tails [21], which is usually the case in the financial time series [5]. Note that the spectrum-based estimators of $H_{xy}$ [16] are not biased by the heavy tails. And third, there is a finite sample bias as showed in detail in [19]. Unfortunately, this bias can be either positive, negative or none depending on the level of correlation between series for scales close to zero. This makes $H_{xy}$ or specifically its comparison with $\frac{1}{2}(H_x + H_y)$ unreliable.

What makes this finding even more alarming is the fact that in the financial econometrics and time series analysis literature, the impossibility of $H_{xy} > \frac{1}{2}(H_x + H_y)$ is taken as an obvious property and it is pretty much a two-liner in [36] who quickly focuses on the $H_{xy} < \frac{1}{2}(H_x + H_y)$ case as the only relevant one for further analysis.

## All in Vain?

One might then ask whether the whole research around the power-law cross-correlations is in vain and futile. The short answer is "no" but it needs further work with more care about theoretical aspects of the topic. As it stands, most of the empirical literature reports either $H_{xy} > \frac{1}{2}(H_x + H_y)$ or $H_{xy} = \frac{1}{2}(H_x + H_y)$. The former is infeasible, i.e. wrong, and the latter is not interesting as it is implied by many different models. In addition, the latter case is simply a reflection of power-law auto-correlations of the separate processes (or at least one of them) and the fact that the processes are pairwise correlated, nothing else is needed for $H_{xy} = \frac{1}{2}(H_x + H_y)$ to hold [20]. There are two ways how this research branch can be further exploited even without $H_{xy}$ and its sole interpretation—utilizing the construction of DCCA without needing $H_{xy}$, and focusing on the case of $H_{xy} < \frac{1}{2}(H_x + H_y)$.

## *Scale-Specific Correlations and Regressions*

The DCCA procedure is built on scaling of the bivariate fluctuation function $F^2_{XY}(s)$, which eventually leads to a power-law scaling $F^2_{XY}(s) \propto s^{2H_{xy}}$, in the same way the DFA procedure is based on the fluctuation function $F^2(s)$ scaling. Asymptotically, these can be seen as covariance and variance, respectively, relative to the specific scale $s$, i.e. a scale-specific covariance $F^2_{XY}(s)$ and a scale-specific variance $F^2(s)$. This idea has been further expanded by [41] who proposed the DCCA-based correlation coefficient as

$$\rho_{DCCA}(s) = \frac{F^2_{XY}(s)}{\sqrt{F^2_X(s)F^2_Y(s)}}$$

where $F^2_X(s)$ and $F^2_Y(s)$ are scale-specific variances of processes $X$ and $Y$. This correlation coefficient has been shown to work well for non-stationary series as well and to outperform the standard Pearson correlation coefficient [14]. In addition, its construction is so straightforward and appealing that it is quite easy to construct such correlation coefficients using almost any power-law cross-correlation methods [15].

When the scale-dependent correlations are defined, it is only a simple step towards regression. Kristoufek [18] introduces a DCCA-based estimator of the scale-dependent $\beta$ coefficient, defined as

$$\hat{\beta}^{DCCA}(s) = \frac{F^2_{XY}(s)}{F^2_X(s)}.$$

Compared to $\rho_{DCCA}(s)$, which measures the strength of the relationship, $\hat{\beta}^{DCCA}(s)$ gives the specific effect, i.e. its level, which is much more useful for interpretation of economic and financial relationships where one is usually interested not only in whether the variables are strongly or weakly correlated but what the actual effect of one variable on another is.

The work and insight of [41] has thus given a very important alternative utility of the DCCA method (and other time domain $H_{xy}$ in general) and he has shown that the in-between steps of methods can sometimes lead to completely novel views on the topic.

### Power-Law Coherency

As noted by [19, 22, 36], only the case of $H_{xy} < \frac{1}{2}(H_x + H_y)$ is an interesting venue as it promises a new class of processes. Returning back to utilizing the squared spectrum coherency, if the two processes are power-law correlated so that $f_x(\omega) \propto \omega^{1-2H_x}$ and $f_y(\omega) \propto \omega^{1-2H_y}$ close to the origin ($\omega \to 0+$) and they are power-law cross-correlated so that $|f_{xy}(\omega)| \propto \omega^{1-2H_{xy}}$ close to the origin, we can write

$$K_{xy}^2(\omega) \propto \frac{\omega^{2(1-2H_{xy})}}{\omega^{1-2H_x}\omega^{1-2H_y}} = \omega^{-4(H_{xy}-\frac{H_x+H_y}{2})} \equiv \omega^{-4H_\rho}$$

close to the origin. The power-law coherency can be defined through parameter $H_\rho$ as $H_\rho = H_{xy} - \frac{H_x+H_y}{2}$.

Recall that the squared spectrum coherency $0 \leq K_{xy}^2(\omega) \leq 1$ for all frequencies $\omega$ which yields only two possible settings for the exponent—either $H_\rho = 0$ or $H_\rho < 0$. $H_\rho = 0$ gives us $H_{xy} = \frac{H_x+H_y}{2}$ and the coherency goes to a constant for very low frequencies $\omega$. The more interesting situation arises when $H_\rho < 0$ (resulting in $H_{xy} < \frac{H_x+H_y}{2}$) which implies that the squared coherency goes to zero for low frequencies approaching zero, specifically in the power-law manner (hence power-law coherency). The power-law coherent processes can be correlated at high frequencies but are uncorrelated at low frequencies. From the perspective of financial economics, these processes can be correlated in the short term but are uncorrelated in the long term (and that is why [36] refer to such processes as anti-cointegration). Such processes have potentially huge impact on portfolio construction and risk management as an asset characterized as such would serve as important risk diversifier from the long-term perspective.

As in detailed shown by [22], the power-law coherency can be translated into the time domain easily. Eventually, one arrives at

$$\rho_{xy}^2(s) \propto \frac{s^{4H_{xy}}}{s^{2H_x}s^{2H_y}} = s^{4(H_{xy}-\frac{H_x-H_y}{2})} \equiv s^{4H_\rho}$$

so that the scaling exponent for both time- ($s \to +\infty$) and frequency- ($\omega \to 0+$) domain power-law coherency is the same. Interestingly, the squared correlation $\rho_{xy}^2(s)$ can be easily represented by the squared DCCA-based correlation coefficient $\rho_{DCCA}^2(s)$ so that both approaches presented in this section and the previous one nicely connect in the end. This parallel view gives another insight into the interpretation of the relationship between $H_{xy}$ and $\frac{1}{2}(H_x + H_y)$, specifically two interesting cases. When $H_{xy} = \frac{H_x+H_y}{2}$ and $\rho_{xy}^2(s) \approx 1$ for $s \to +\infty$, we have possible cointegration, i.e. the variables are not necessarily connected in the short term but are tightly connected in the long term. And when $H_{xy} < \frac{H_x+H_y}{2}$, we have anti-cointegration, i.e. the variables are not related in the long term but are possibly connected in the short term.

## Discussion

The whole issue and suggested solutions presented above point mainly to a general problem of the interdisciplinary research (here specifically econophysics)—communities do not interact enough. The bivariate breakthrough in the sense of power-law cross-correlations came in 2008 [33] but the interpretation of the bivariate cross-correlations parameter was already obviously problematic. Already in 2009, [35] discussed the multivariate fractionally integrated processes (power-law cross-correlated in the econophysics language) and took as a given that the bivariate cross-correlation parameter is practically forced by the autocorrelation properties of the separate processes. This was only further extended in [36] but the crucial result has been there since 2009. Moreover, the topic of bivariate dependence in the long-range correlations setting was not new and it had been discussed in the financial econometrics literature much earlier [23, 24] and already [24] discusses the connection between the separate and bivariate fractional integration parameters (equivalent to the long-range dependence in the financial econometrics language). Only it has not come to the econophysics community and it had to be re-discovered first by [33], who introduced the DCCA estimator of the $H_{xy}$ parameter, and [19], who showed that $H_{xy} > \frac{H_x+H_y}{2}$ is impossible and only $H_{xy} < \frac{H_x+H_y}{2}$ is theoretically and practically appealing, and translated the framework into the more standard (for econophysics) time-domain language.

However, both directions of communication can be beneficial. Financial econometrics focuses a lot on the theoretical properties of estimators as well as their restrictions and assumptions. As shown in the text above, the power-law cross-correlations literature has missed a lot of it since the very beginning. The econophysics community can learn from it and see that quite often, there is a need for theoretical background before jumping into empirical avenues. From the other side, financial econometrics often focuses on assumptions and restrictions too much, missing a big picture and possible usefulness of various methods. The time-domain approaches that are so characteristic for the econophysics field are usually omitted in the financial econometrics literature because it is more complicated to show their statistical properties compared to the frequency-domain approaches. However, when the methods from both domains are compared under different statistical assumptions, the time-domain methods often prevail (compare the results in [16, 19, 21, 22]).

With the increasing computational power, the non-existence (or negligence) of the asymptotic properties for the econophysics methods is becoming less of a problem as the properties can be simulated even for very long time series with various dynamic properties. As it turns out, many of these econophysics methods can compete and even beat the frequency-domain methods, which are popular in the economics and econometrics mainstream, even for very long time series which would standardly be considered as a good enough approximation of infinity, i.e. asymptotics. Econophysics has always boasted with being data-driven, empirically based science discipline, which has certainly led to many breakthroughs. However, and this is specifically true for the last couple of years, the econophysics literature has been flooded with empirical papers that are simple analyses of the "choose the method, input the data, list the results" type without much effort of results interpretation. For econophysics to be treated with more respect by the mainstream economics and finance communities, the concrete practical implications and applications, such as specific policy suggestions, trading strategies, portfolio methods and similar, need to be presented. The same issue has been the case for the power-law cross-correlations since the very beginning outlined by [33]. There has been no interpretation or practical implications of the bivariate Hurst exponent $H_{xy}$. It has been standardly stated that the two series are "power-law cross-correlated" or "cross-persistent" with no hint of what it actually implies (in a practical sense) for the series dynamics.

In this text, it has been shown that the power-law cross-correlations setting is inherently problematic and the bivariate Hurst exponent alone does not give any information about the relationship between two analysed series. Only a comparison of $H_{xy}$ with its separate counterparts gives any information. However, unless the series are pairwise independent (or at least uncorrelated) or they are the unique case of anti-cointegration, it will automatically hold that $H_{xy} = \frac{H_x+H_y}{2}$ which, unfortunately, covers a wide range of very different possible relationships between the series so that its informative value is again miniature. Fortunately, there are at least two ways (recalled in this text) of utilizing partial steps and results of the power-law cross-correlations setting—scale-specific correlations and regressions, and power-law coherency. Nevertheless, the challenges for these two approaches remain the same and should not be overshadowed by purely empirical studies—to show practical utility of the methods. For each of these, the utility seems at hand—portfolio construction. This is the case both for the scale-specific correlations and regression as these can be used as standard correlation matrices or $\beta$-parameters in the capital asset pricing model setting, and for the anti-cointegration case which promises high-quality long-term diversifiers. Only then will we be allowed to say that power-law correlations have been contributive and useful.

**Acknowledgements** Ladislav Kristoufek gratefully acknowledges financial support of the Czech Science Foundation (project 17-12386Y).

## References

1. J. Barunik, L. Kristoufek, On Hurst exponent estimation under heavy-tailed distributions. Physica A **389**(18), 3844–3855 (2010)
2. J. Beran, *Statistics for Long-Memory Processes*. Monographs on Statistics and Applied Probability, vol. 61 (Chapman and Hall, New York, 1994)
3. T. Bollerslev, Generalized autoregressive conditional heteroskedasticity. J. Econ. **31**(3), 307–327 (1986)
4. G. Box, G.M. Jenkins, G.C. Reinsel, *Time Series Analysis: Forecasting and Control* (Prentice-Hall, Upper Saddle River, 1994)
5. R. Cont, Empirical properties of asset returns: stylized facts and statistical issues. Quant. Financ. **1**(2), 223–236 (2001)

6. R.F. Engle, Autoregressive conditional heteroskedasticity with estimates of variance of United Kingdom inflation. Econometrica **50**(4), 987–1007 (1982)
7. D. Grech, Z. Mazur, Statistical properties of old and new techniques in detrended analysis of time series. Acta Phys. Pol. B **36**, 2403–2413 (2005)
8. L.-Y. He, S.-P. Chen, A new approach to quantify power-law cross-correlation and its application to commodity markets. Physica A **390**, 3806–3814 (2011)
9. H. Hurst, Long term storage capacity of reservoirs. Trans. Am. Soc. Eng. **116**, 770–799 (1951)
10. J. Kantelhardt, S. Zschiegner, E. Koscielny-Bunde, A. Bunde, S. Havlin, E. Stanley, Multifractal detrended fluctuation analysis of nonstationary time series. Physica A **316**(1–4), 87–114 (2002)
11. L. Kristoufek, Rescaled range analysis and detrended fluctuation analysis: finite sample properties and confidence intervals. AUCO Czech Econ. Rev. **4**, 236–250 (2010)
12. L. Kristoufek, Mixed-correlated ARFIMA processes for power-law cross-correlations. Physica A **392**, 6484–6493 (2013)
13. L. Kristoufek, Testing power-law cross-correlations: Rescaled covariance test. Eur. Phys. J. B **86** (2013). Art. 418
14. L. Kristoufek, Detrending moving-average cross-correlation coefficient: measuring cross-correlations between non-stationary series. Physica A **406**, 169–175 (2014)
15. L. Kristoufek, Measuring correlations between non-stationary series with DCCA coefficient. Physica A **402**, 291–298 (2014)
16. L. Kristoufek, Spectrum-based estimators of the bivariate Hurst exponent. Phys. Rev. E **90**, 062802 (2014)
17. L. Kristoufek, Can the bivariate Hurst exponent be higher than an average of the separate Hurst exponents? Physica A **431**, 124–127 (2015)
18. L. Kristoufek, Detrended fluctuation analysis as a regression framework: estimating dependence at different scales. Phys. Rev. E **91**, 022802 (2015)
19. L. Kristoufek, Finite sample properties of power-law cross-correlations estimators. Physica A **491**, 513–525 (2015)
20. L. Kristoufek, On the interplay between short- and long-term memory in the power-law cross-correlations setting. Physica A **421**, 218–222 (2015)
21. L. Kristoufek, Power-law cross-correlations estimation under heavy tails. Commun. Nonlinear Sci. Numer. Simul. **40**, 163–172 (2016)
22. L. Kristoufek, Fractal approach towards power-law coherency to measure cross-correlations between series. Commun. Nonlinear Sci. Numer. Simul. **50**, 193–200 (2017)
23. I. Lobato, Consistency of the average cross-periodogram in long memory time series. J. Time Ser. Anal. **18**, 137–155 (1997)
24. I. Lobato, A semiparametric two-step estimator in a multivariate long memory model. J. Econ. **90**, 129–153 (1999)
25. B. Mandelbrot, The variation of some other speculative prices. J. Bus. **40**(4), 393–413 (1967)
26. B. Mandelbrot, J. van Ness, Fractional Brownian motions, fractional noises and applications. SIAM Rev. **10**(422), 422–437 (1968)
27. B. Mandelbrot, J. Wallis, Noah, Joseph and operational hydrology. Water Resour. Res. **4**, 909–918 (1968)
28. R.N. Mantegna, H.E. Stanley, *An Introduction to Econophysics* (Cambridge University Press, Cambridge, 2000)
29. P. Oswiecimka, S. Drozdz, M. Forczek, S. Jadach, J. Kwapien, Detrended cross-correlation analysis consistently extended to multifractality. Phys. Rev. E **89**, 023305 (2014)
30. C. Peng, S. Buldyrev, A. Goldberger, S. Havlin, M. Simons, H. Stanley, Finite-size effects on long-range correlations: implications for analyzing DNA sequences. Phys. Rev. E **47**(5), 3730–3733 (1993)
31. C. Peng, S. Buldyrev, S. Havlin, M. Simons, H. Stanley, A. Goldberger, Mosaic organization of DNA nucleotides. Phys. Rev. E **49**(2), 1685–1689 (1994)
32. B. Podobnik, D. Horvatic, A. Lam Ng, H. Stanley, P. Ivanov, Modeling long-range cross-correlations in two-component ARFIMA and FIARCH processes. Physica A **387**, 3954–3959 (2008)

33. B. Podobnik, H. Stanley, Detrended cross-correlation analysis: a new method for analyzing two nonstationary time series. Phys. Rev. Lett. **100**, 084102 (2008)
34. G. Samorodnitsky, Long range dependence. Found. Trends® Stoch. Syst. **1**(3), 163–257 (2006)
35. R. Sela, C. Hurvich, Computationally efficient methods for two multivariate fractionally integrated models. J. Time Ser. Anal. **30**(6), 631–651 (2009)
36. R. Sela, C. Hurvich, The average periodogram estimator for a power law in coherency. J. Time Ser. Anal. **33**, 340–363 (2012)
37. M. Taqqu, W. Teverosky, W. Willinger, Estimators for long-range dependence: an empirical study. Fractals **3**(4), 785–798 (1995)
38. M. Taqqu, V. Teverovsky, On estimating the intensity of long-range dependence in finite and infinite variance time series, in *A Practical Guide To Heavy Tails: Statistical Techniques and Applications* (1996)
39. V. Teverovsky, M. Taqqu, W. Willinger, A critical look at Lo's modified r/s statistic. J. Stat. Plan. Inference **80**(1–2), 211–227 (1999)
40. F. Wang, G.-P. Liao, J.-H. Li, R.-B. Zou, W. Shi, Cross-correlation detection and analysis for California's electricity market based on analogous multifractal analysis. Chaos **23**, 013129 (2013)
41. G. Zebende, DCCA cross-correlation coefficient: quantifying level of cross-correlation. Physica A **390**, 614–618 (2011)

# Chapter 4
# Multi-phase Long-Term Autocorrelated Diffusion: Stationary Continuous-Time Weierstrass Walk Versus Flight

**Tomasz Gubiec, Jarosław Klamut, and Ryszard Kutner**

**Abstract** We examine diffusion properties of the stationary hierarchical continuous-time Weierstrass walk (CTWW). We show it is a multi-phase representation of the Lévy walk. The hierarchical spatial-temporal coupling together with coupling between dynamic variables defines the CTWW process. The walker moves here with a piecewise constant velocity between trajectory turning points. We found the diffusion phase diagram of the CTWW consisting of (i) the Brownian yet non-Gaussian phase and (ii) the anomalous non-Gaussian and/or not-fBm phases. We compare the diffusion phase diagram of the stationary CTWW with the corresponding stationary hierarchical continuous-time Weierstrass flight (CTWF). The instantaneous jumps between trajectory turning points preceded by waiting define the CTWF process. It is a hierarchical representation of the Lévy flight. We found the diffusion phase diagram of the CTWF as a small part of the corresponding CTWW one.

## Historical Sketch and Inspirations

### *Initial Remarks*

By the pioneering work published in the year 1965 [1], physicists Eliott W. Montroll and George H. Weiss introduced the concept of continuous-time random walk (CTRW) as a way to achieve the interevent-time continuous and fluctuating.[1] This

[1]Let us incidentally comment that term "walk" in the name "continuous-time random walk" is commonly used in the generic sense comprising two concepts, namely, both the walk (associated with the finite velocity of the process) and flight (associated with an instantaneous displacement of the process). Thus, it has to be specified in concrete consideration what kind of a process we have.

Partially supported by the Center for Study of Systemic Risk, Faculty of "Artes Liberales", University of Warsaw.

T. Gubiec (✉) · J. Klamut · R. Kutner (✉)
Faculty of Physics, University of Warsaw, Pasteur Str. 5, 02093 Warsaw, Poland
e-mail: ryszard.kutner@fuw.edu.pl



D. Grech and J. Miśkiewicz (eds.), *Simplicity of Complexity in Economic and Social Systems*, Springer Proceedings in Complexity,
https://doi.org/10.1007/978-3-030-56160-4_4

stochastic process characterizes by the waiting-time and displacement distribution[2] (WTDD), giving an insight into the process details in various scales. The WTDD permits the description of both the Debye (exponential) and non-Debye (slowly decaying) relaxations as well as regular and anomalous transport and diffusion [2, 3]. It is promising that the WTDD is also used to characterize evolving complex systems [4].

The publication [1] mentioned above stirred up a little attention of a community, until in the early 70s with the works of Harvey Scher and collaborators [5]. They have applied the CTRW technique to the anomalous transport and diffusion in amorphous materials. The article of Shlesinger [6] presents a thorough review of the history and antecedents of the CTRW.

In subsequent years, the importance of the formalism soared with an increasing number of works generalizing and applying the CTRW with the number of cites climbing to several tens of thousands. The CTRW constitutes an extremely powerful, albeit relatively simple formalism to approach and eventually solve a countless number of problems in many areas of physical, natural and even socio-economic sciences. In this regard, see our review article [7].

Presumably, the most commonly used CTRW formalism was developed by Scher and Lax in terms of recursion relations [8–13]. In this context, the distinction between separable and non-separable WTDDs, key for present work, was introduced [14]. A thorough analysis of the non-separable one [15–17] took into account dependence over many correlated consecutive particle displacements and waiting (or inter-event) times.

The canonical version of the CTRW formalism, concerning transitions between different sites and states by using the recursion relations, is equivalent with generalized master equation (GME). This is thanks to the one-to-one transformation between WTDD and memory kernel was clearly established [18–26]. This creates the foundation of our work because memory can lead to the anomalous or/and non-Gaussian diffusion.

Although, initially, the CTRW formalism was a kind of renewal theory, Tunaley was able to modify it by preparing the class of initial (averaging) WTDD, crucial for this work. Such a modification makes time homogeneous [27, 28] and enables to consider CTRW as the semi-Markov process [29, 30]. Thus, the application of the fundamental Wiener-Khinchin theorem (relating autocorrelation function to power spectra) became possible.

In principle, the CTRW is fundamentally different than the regular random flight or walk models. As part of CTRW formalism, its WTDD can scale in the long-time (asymptotic) limit in a non-Gaussian way. It is a severe and inspirational extension of the Gaussian scaling. Thus, the CTRW became a foundation of anomalous (dispersive, non-Gaussian) transport and diffusion [13, 31, 32] opening the modern and trendy segment of statistical physics, as well as condensed and soft matter

[2]This distribution can be also called pausing- or interevent time and displacement distribution. Previously, the incomplete name "waiting-time distribution" (WTD) was used.

physics. Moreover, the CTRW formalism rapidly expanded even outside the traditional physics [33–35]. In this work, we consider multi-phase diffusion by analysing diffusion phase diagrams.

## *Elements of Lévy Random Motion*

Thanks to its versatility, the CTRW found numerous vital applications in many fields ranging from biology through telecommunication to finance including econometrics and economics, and even to speech recognition. The CTRW found countless applications in many other areas, still growing, such as the ageing of glasses, a nearly constant dielectric loss response in structurally disordered ionic conductors as well as in modelling of hydrological models and earthquakes.

When particle performs CTRW of the flight (jump or hop) type then during its evolution, it makes instantaneous jumps alternated with waiting events (or rests). The CTRW formalism enables to combine both particle states, offering an abundant diffusion phase diagram or several scaling regimes. Moreover, the CTRW formalism can be extended assuming walks with finite velocity instead of the instantaneous jumps. Such a kind of model we call the Lévy walk interrupted by rests [3, 36–38]. The presence of finite particle velocity there significantly increases the flexibility of this kind of models. However, it simultaneously makes it more difficult a finding of their analytical solutions if they exist. The standard versions of Lévy walk model, i.e. without rests, were also intensively developed assuming fixed particle velocity [39, 40] or varying, e.g. according to the self-similar hierarchical structure [41]. The present work is its extension [41].

There are several generalizations of the Lévy walk model [3] assuming that particle velocity can vary randomly [42, 43] or by some other rules [43]. Among them, Lévy walk model with random velocity (LWRV) and the one with weakly fluctuating velocity (LWWFV) caused by the active environment [44–46] are inspiring. The LWRV model was applied, for instance, (i) in physical problems of two-dimensional turbulence [47–49], (ii) modelling of velocity distribution in kinetic theory [50, 51], (iii) in plasma physics [52, 53] and (iv) in some non-extensive statistics [54, 55].

The finite velocity of walking particles constitutes random walk models more general than the erratic flight (jump) ones. It brings them closer to the physical principles. Indeed, more fluctuations are responsible for real-life aspects, which CTWW formalism takes into account in contrast to CTWF one. These two formalisms are systematic compared with each other in this work giving a contribution to a stream of works on the subject [38, 56–58].

## Stationary Continuous-Time Weierstrass Walk Versus Flight: Definitions

When we consider the continuous-time Weierstrass walk we allow the walker to move with a constant velocity between the successive trajectory turning points [41, 59, 60]. That is, the walker (process) velocity is a piecewise constant function of time. Figure 4.1 shows a comparison of typical CTWF and CTWW trajectories' segments.

Both processes (CTWF and CTWW) we define by (2) and (3). The location of turning points of the CTWF is defined utilizing instantaneous single-step displacements $x = \pm b_0 b^j = \pm v_0 v^j \langle \Delta t_j \rangle$, where $\langle \Delta t_j \rangle = \tau_0 \tau^j$, $j = 0, 1, \ldots, b_0 = v_0 \tau_0, b = v\tau$. In the case of the CTWW, the corresponding turning points are defined using the temporary single-step displacements $x = \pm v_0 v^j \Delta t_j$ (discussed in Appendix "CTWW Propagator"), while the stochastic variable $\Delta t_j = \Delta t$, where $\Delta t$ is in each step drawn from the stationary exponential distribution characterized by the relaxation time $\tau_0 \tau^j$.

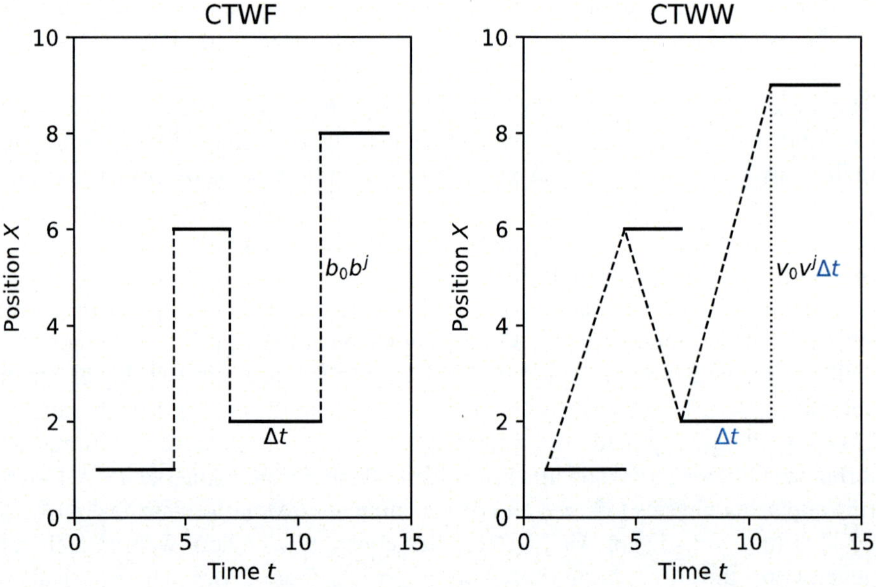


**Fig. 4.1** The comparison of typical one-dimensional trajectories of the CTWF (left plot) and CTWW (right plot) processes. The horizontal fluctuating time intervals $\Delta t$ (substantial horizontal segments) denote process time lags between successive turning points. In the case of the CTWF process, it means waiting or pausing time. Whereas, in the case of the CTWW process, it means the interevent time needed to overcome segments $\pm v_0 v^j \Delta t$ of the trajectory between successive turning points. In the case of CTWF, parts $\pm b_0 b^j$ of the path are travelled abruptly (dashed vertical segments of straight lines). Whereas, in the case of the CTWW, the dashed sloped fragments of straight lines mimic the walk at a constant velocity of $v_0 v^j$. Recall that both $j$ and $\Delta t$ are random variables (discussed in section "Stationary Continuous-Time Weierstrass Walk Versus Flight: Definitions"), which can be different for each pair of successive turning points

For both processes, the index $j$ of the hierarchy's level is drawn from the geometric distribution,

$$w(j) = \left(1 - \frac{1}{N}\right)\frac{1}{N^j}. \quad (1)$$

We obtained this distribution simply assuming that the ratio $\frac{w(j+1)}{w(j)}$ is stationary, i.e. assuming its independence from index $j$.

Before we move on to propagator's construction in section "Propagator", we clarify the general definitions of both types of processes using single-step distributions. Figure 4.2 is helpful here.

These are generic definitions independent of a particular form of distribution. They base on waiting-time and displacement joint distributions (WTDDs). These distributions differ both in the scaling of space variable and physical interpretations. However, we wanted this difference to be as small as possible and still leading to fundamentally different phase diagrams in a long-time limit. In general, we can write

$$\psi^F(x, \Delta t) = \sum_{j=0}^{\infty} w(j)\frac{1}{b_0 b^j}\frac{1}{\tau_0 \tau^j} \times \phi\left(\frac{x}{b_0 b^j}, \frac{\Delta t}{\tau_0 \tau^j}\right), \text{ for CTWF} \quad (2)$$

and

$$\psi^W(x, \Delta t) = \sum_{j=0}^{\infty} w(j)\frac{1}{v_0 v^j \Delta t}\frac{1}{\tau_0 \tau^j} \times \phi\left(\frac{x}{v_0 v^j \Delta t}, \frac{\Delta t}{\tau_0 \tau^j}\right), \text{ for CTWW} \quad (3)$$

where index $j$ defines the level of the stochastic hierarchy and $j$-independent scaling function $\phi(y, \vartheta)$ is normalized, that is, $\int_0^\infty d\vartheta \int_{-\infty}^\infty dy \phi(y, \vartheta) = 1$. Besides, we assume that the scaling function $\phi$ is symmetric in its spatial argument $y$. We emphasize that the form of the scaling function $\phi$ is the same for CTWF and CTWW and differs only in the scaling of the spatial variable $x$. In the case of CTWF, this variable equals $b_0 b^j$, while in the case of CTWW, it is $v_0 v^j \Delta t$. We continue to use indexes "F" and "W" for the CTWF and CTWW processes, respectively.

What's more, we can now write conditional (or partial) WTDD assigned to the individual hierarchy level $j = 0, 1, 2, \ldots$, in the form

$$\psi_j^F\left(\frac{x}{b_0 b^j}, \frac{\Delta t}{\tau_0 \tau^j}\right) = \frac{1}{b_0 b^j}\frac{1}{\tau_0 \tau^j} \times \phi\left(\frac{x}{b_0 b^j}, \frac{\Delta t}{\tau_0 \tau^j}\right), \text{ for CTWF} \quad (4)$$

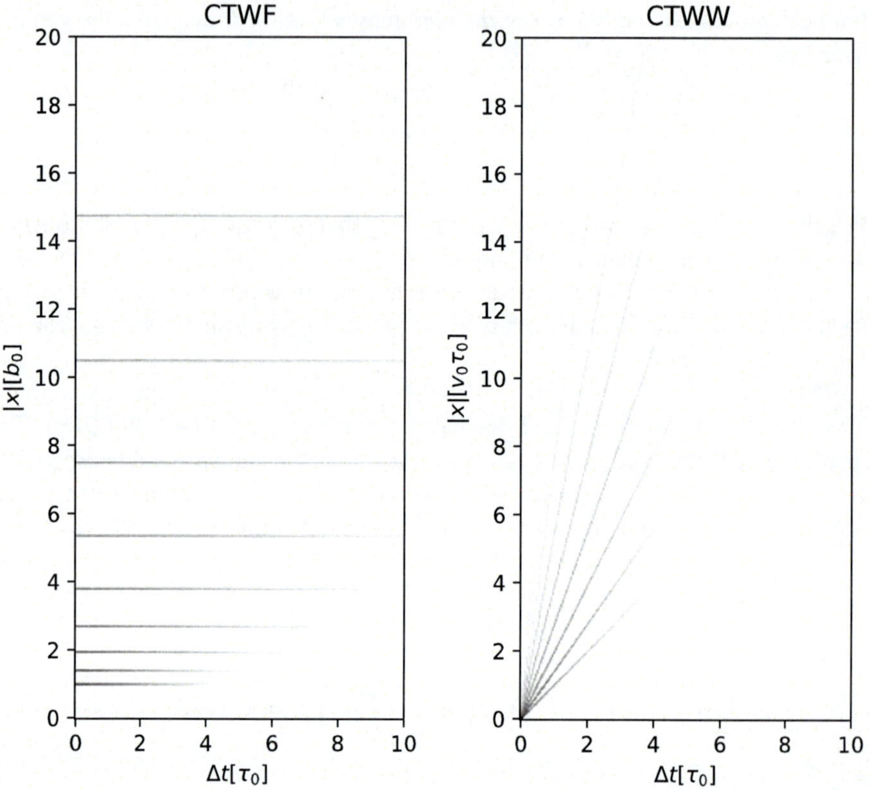


**Fig. 4.2** The comparison of waiting-time and displacement distributions, WTDDs, for the CTWF (left plot) and CTWW (right plot) processes based on their definitions (2) and (3), respectively. These definitions are supported herein by definition (14). We show heatmaps in resolution 200px times 400px and integrate WTDDs over the area of each pixel. The different shades of grey of each pixel correspond to the different values of integrated WTDD. The greyness is brighter, the lower this value. Because for the CTWF, displacements $\mid x \mid= b_0 b^j,\ \ j = 0, 1, 2, \ldots$ are independent of $\Delta t$, they are shown as the horizontal lines. For CTWW, analogous relationships take the form $\mid x \mid= v_0 v^j \Delta t$, i.e. they linearly depend on $\Delta t,\ \ j = 0, 1, 2, \ldots$. That is why we present them in the form of diagonal straight lines. These horizontal and diagonal lines are numbered by index $j$ sequentially from the bottom lines. Going up on both plots we move to straight lines with increasing brightness of grey. This brightness increases also along the individual straight lines according to the exponential distribution $p_j(\Delta t) = \frac{1}{\tau_0 \tau^j} \exp\left(-\Delta t/\tau_0 \tau^j\right)$ weighted by (1). For both plots, we assume, for example, $b = 1.4$, $v = 1.4$, $\tau = 1.3$ and $N = 1.05$

and

$$\psi_j^W\left(\frac{x}{v_0 v^j \Delta t}, \frac{\Delta t}{\tau_0 \tau^j}\right) = \frac{1}{v_0 v^j \Delta t} \frac{1}{\tau_0 \tau^j} \times \phi\left(\frac{x}{v_0 v^j \Delta t}, \frac{\Delta t}{\tau_0 \tau^j}\right), \text{ for CTWW} \tag{5}$$

which makes their physical interpretation easier.

That is, $\psi_j^F$ is the density of the conditional probability that the process is waiting at some turning point (for exactly $\Delta t$ duration time) and then executes hopping by displacement $x$ to another turning point. The mentioned condition means that this $\psi_j^F$ we assign to the $j$th level of the hierarchy. However, $\psi_j^W$, although related to the $j$th level of the hierarchy, has a different physical interpretation. It is the density of the conditional probability that the process runs between successive turning points with a constant $j$-dependent velocity $v_0 v^j$ precisely for the duration of $\Delta t$.

The role of index $j$ is to prioritize both displacements and the corresponding times. It is a coarse-grained approach consisting of grouping these dynamic variables according to the power of the basis $b$ and $\tau$, respectively. The power of $N$ also weights the levels of the hierarchy constructed in this way.

Notably, in the CTWW process, we are dealing with an additional coupling between single-step displacement and $\Delta t$. It is this coupling that is crucial to distinguish between the two processes discussed—this we illustrate in Fig. 4.2.

### *Useful Relations Between Single-Step Moments*

Thanks to the definitions above, we can show what the difference between CTWW and CTWF processes is. We calculate for this purpose variances of the single-step displacements at the $j$th level of the hierarchy, separately for both processes. Using (4) we get variance at level $j$ for the CTWF in the simple form,

$$\langle x^2 \rangle_j^F = b_0^2 b^{2j} \langle y^2 \rangle, \tag{6}$$

where $\langle y^2 \rangle = \int_0^\infty d\vartheta \int_{-\infty}^\infty dy y^2 \phi(y, \vartheta)$ is the marginal second moment of the scale-free space variable. Similarly, using (5), we have

$$\langle x^2 \rangle_j^W = b_0^2 b^{2j} \langle y^2 \vartheta^2 \rangle, \tag{7}$$

where $\langle y^2 \vartheta^2 \rangle = \int_0^\infty d\vartheta \vartheta^2 \int_{-\infty}^\infty dy y^2 \phi(y, \vartheta)$.

From (6) and (7), we obtain

$$\langle x^2 \rangle_j^W = \langle x^2 \rangle_j^F \frac{\langle y^2 \vartheta^2 \rangle}{\langle y^2 \rangle}. \tag{8}$$

For variables $x$ and $\Delta t$ separable (in a multiplicative form) for any level of the hierarchy, i.e. obeying equality $\langle y^2 \vartheta^2 \rangle = \langle y^2 \rangle \langle \vartheta^2 \rangle$, (8) simplifies to the form

$$\langle x^2 \rangle_j^W = \langle x^2 \rangle_j^F \langle \vartheta^2 \rangle, \tag{9}$$

where $\langle \vartheta^2 \rangle = \int_0^\infty d\vartheta \vartheta^2 \int_{-\infty}^\infty dy \phi(y, \vartheta)$.

In this work, we consider solely separability considered above. This separability applies to each level $j$ of the hierarchy separately. This is due to the factored form of the scaling function $\phi$ adopted later by (14). It is clear that the difference between single-step space variances for both processes comes only from the variance of the random variable $\Delta t$ as $\langle\vartheta^2\rangle > \langle\vartheta\rangle^2$, i.e. from the non-vanishing variance $\sigma_{\vartheta}^2 = \langle\vartheta^2\rangle - \langle\vartheta\rangle^2$. Of course, this variance depends in a significant way on the probability distribution $\phi$ used here to prepare the average values, but it is independent of the hierarchy.

Analogous expressions we also have for global values, i.e. averaged overall levels of the hierarchy. Namely,

$$\langle x^2\rangle^W = \langle x^2\rangle^F \frac{\langle y^2\vartheta^2\rangle}{\langle y^2\rangle} = b_0^2 Y \langle y^2\vartheta^2\rangle, \tag{10}$$

corresponding to (8) and

$$\langle x^2\rangle^W = \langle x^2\rangle^F \langle\vartheta^2\rangle = b_0^2 Y \langle y^2\rangle\langle\vartheta^2\rangle, \tag{11}$$

corresponding to (9), where $\langle x^2\rangle^{F,W} = \sum_{j=0}^{\infty} w(j)\langle x^2\rangle_j^{F,W}$ and

$$Y = \begin{cases} \frac{1-\frac{1}{N}}{1-\frac{b^2}{N}}, & \text{for } \beta > 2, \\ \infty, & \text{for } \beta < 2, \end{cases} \tag{12}$$

both for the CTWF and CTWW processes.

The first equality in (11) we can rewrite in a separated additive form accessible to the physical interpretation,

$$\langle x^2\rangle^W = \langle x^2\rangle^F \sigma_{\vartheta}^2 + \langle x^2\rangle^F \langle\vartheta\rangle^2, \tag{13}$$

where the first component concerns fluctuation of $\theta$ variable, while the second component concerns its drift. Let us emphasize that the fluctuation component exists because the unbalanced fluctuations occur in the time-symmetry breaking situation. That is, we assumed that the arrow of time works herein. As one can see the difference between CTWW and CTWF already at the level of a single displacement is significant.

We derive expressions for arbitrary even moments (stationary and non-stationary) in Appendix "Non-factorial Spatial-Temporal Cross-Moments" for completeness.

We assume the scaling function $\phi$ in the following simple factored form:

$$\phi(y,\theta) = \frac{1}{2}\left[\delta(y-1) + \delta(y+1)\right]\exp(-\theta). \tag{14}$$

This scaling function already suffice to present the fundamental differences between CTWF and CTWW, what we show in Fig. 4.2.

### *Main Subject*

The situation for multi-step variances is much more complicated due to the possibility of cumulative fluctuations. It leads to a significant distinction between CTWF and CTWW processes in a long time, both on the level of regular (Brownian) and singular (anomalous) behaviours. Explanation of this distinction is the main subject of this work. However, we must remember that the CTWW process only mimics the walk process.

We prove that CTWW provides an abundant diffusion phase diagram even in a stationary situation, consisting of several correctly classified phases. The stationary situation means here the independence of the random walk process from choosing the starting point of this process. We show in section "Mean-Square Displacement and Autocorrelation Function in a Closed Form" that stationary diffusion phase diagrams that we build are the result of regular diffusion competition (also called linear or Brownian diffusion) with anomalous diffusion. What's more, each phase (even the regular one) is characterized by the autocorrelation function, which disappears according to power law with some combined exponent. In the further part of the work, we compare phase diagrams for CTWW and CTWF. Besides, we calculate kurtosis to check if we are dealing with fractional Brownian motion (fBm) at any phase of the diagram. Finally, we set the explicit form of the propagator. We emphasize that both CTWW and CTWF are significant cases of more general CTRW formalism (see section "Basic Distributions and Means" for details).

The construction and proper description of the stationary phase diagrams require the determination of moments and autocorrelation functions. To do this, we must first build a propagator $P_{st}(X, t)$. We define it as the probability density of finding a walker at a position $X$ at time $t$ at condition that time origin can be chosen arbitrarily in the stationary situation. However, we do not require that the propagator builds in a closed form. In this work, we consider the even moment expansion of the propagator as odd moments vanish herein (because no drift is assumed). It allows us to answer the critical question in which conditions we are dealing with Gaussian walk and in which ones with non-Gaussian one.

## Propagator

### *General Stationary Situation*

We use the following expansions of the propagator (see Appendix "Expansions of Propagator" for details). The short, three-term expansion,

$$\tilde{P}_{st}(k,t) = 1 - \frac{1}{2}k^2\langle X(t)^2\rangle_{st} + \tilde{R}^4_{st}(k,t), \tag{15}$$

is used for the phases where the fourth moment $\langle X(t)^4\rangle_{st}$ doesn't exist (i.e. $\langle X(t)^4\rangle_{st} = \infty$). Everywhere in this work the index $st$ is present for the stationary case. Notably, the propagator tends to the Gaussian for vanishing $k$ as then the rest $\tilde{R}^4_{st}(k,t)$ vanishes (see (43) in Appendix "Expansions of Propagator" for details).

The longer, four-term expansion,

$$\tilde{P}_{st}(k,t) = 1 - \frac{1}{2}k^2\langle X(t)^2\rangle_{st} + \frac{1}{4!}k^4\langle X(t)^4\rangle_{st} + \tilde{R}^6_{st}(k,t), \tag{16}$$

we have for phases where the fourth moment exists. In this case, we deal with Gaussian for vanishing $k$ when excess kurtosis (or the fourth cumulant) disappears. That is,

$$\kappa_{st}(t) = \langle X(t)^4\rangle_{st} - 3\langle X(t)^2\rangle^2_{st}, \tag{17}$$

and the rest $\tilde{R}^6_{st}(k,t)$ also vanishes (see (44) in Appendix "Expansions of Propagator" for detail). This is equivalent to the situation of vanishing kurtosis and all higher cumulants if they exist.

In general, we can write

$$\tilde{P}_{st}(k,t) = \sum_{j=0}^{m} \frac{(-1)^j}{(2j)!}k^{2j}\langle X(t)^{2j}\rangle + \tilde{R}^{2(m+1)}_{st}(k,t), \tag{18}$$

for phases where the $(2m)$th moment exists (i.e. $\langle X(t)^{2m}\rangle < \infty,\ m = 0, 1, 2, \ldots$).

Notably, in the case of anomalous diffusion at vanishing $\kappa_{st}(t)$ and $\tilde{R}^6_{st}(k,t)$, we deal with the fractional Brownian motion (fBm). However, the most interesting is the phase beyond the fBm. It includes the case of the Brownian but non-Gaussian diffusion [59, 61]. This phase splits into semi-regular diffusion phases (SRD1 and SRD2) and regular diffusion (RD) phase (see section "Mean-Square Displacement and Autocorrelation Function in a Closed Form" and especially Fig. 4.3 for details).

## *Stationary Propagator and Even Moments: General Representation*

We have (within the stationary CTRW) the following expression for the propagator [41]:

$$\hat{\tilde{P}}_{st}(k,u) = \hat{\tilde{\Xi}}(k,u) + \hat{\tilde{\chi}}(k,u)\hat{\tilde{P}}(k,u), \tag{19}$$

where $\hat{\tilde{F}}$ denotes the Fourier-Laplace transform of $F$ and

$$\hat{\tilde{P}}(k,u) = \frac{\hat{\tilde{\Psi}}(k,u)}{1-\hat{\tilde{\psi}}(k,u)} \tag{20}$$

is the propagator of the non-stationary continuous-time Weierstrass walk [60]. The basic quantities $\hat{\tilde{\psi}}(k,u), \hat{\tilde{\Psi}}(k,u), \hat{\tilde{\chi}}(k,u)$ and $\hat{\tilde{\Xi}}(k,u)$ are defined in Appendix "Basic Distributions and Means".

From (16), (19) and (47) (in Appendix "Expansions of Propagator"), we get

$$\begin{aligned} \langle X(t)^2\rangle_{st} &= -\frac{d^2}{dk^2}\tilde{P}_{st}(k,t)|_{k=0} \\ &\Leftrightarrow \langle \hat{X}(u)^2\rangle_{st} = \langle \hat{X}(u)^2\rangle \\ &- \frac{d^2}{dk^2}\hat{\tilde{\Xi}}(k,u)|_{k=0} - \frac{1}{u}\frac{d^2}{dk^2}\hat{\tilde{\chi}}(k,u)|_{k=0} \end{aligned} \tag{21}$$

and

$$\begin{aligned} \langle \hat{X}(u)^4\rangle_{st} &= \langle \hat{X}(u)^4\rangle \\ &+ \frac{d^4}{dk^4}\hat{\tilde{\Xi}}(k,u)|_{k=0} + \frac{1}{u}\frac{d^4}{dk^4}\hat{\tilde{\chi}}(k,u)|_{k=0} \\ &- 3!\langle \hat{X}(u)^2\rangle\frac{d^2}{dk^2}\hat{\tilde{\chi}}(k,u)|_{k=0} \end{aligned} \tag{22}$$

for the long-time limit (or $| u\tau_0 | \ll 1$), where $\tau_0$ is a time unit. Our calculations (carried out later) indicate that the last component in (22) does not give any contribution to $\langle \hat{X}(u)^4\rangle_{st}$ for the long-time limit.

In the definition (69) of the stationary WTDD (or $\chi(t)$), there is an expected value for the non-stationary state $\langle \Delta t\rangle$. Thus, the transition to a stationary state requires this expected value. It imposes a natural limitation: there must be an exponent $\alpha > 1$ if the stationary state is to exist. We set up converging of microscale or converging a mean waiting time in the non-stationary state.

To better characterize the stationary diffusion phase diagram, we introduced an average waiting time in the stationary state $\langle \Delta t\rangle_{st}$ given by (85) in Appendix "Stationary Case for $n = 0, 1, 2, \ldots$ and $m = 0, 1, 2, \ldots$". This average diverges for $\alpha < 2$, while converges otherwise (marginal case $\alpha = 2$ is not consider herein). Thus, the stationary diffusion phase diagram is divided into two separate parts. The first part for $\frac{1}{\alpha} > \frac{1}{2}$, where we deal with a weak ergodicity breaking [62–70] and the second part for $\frac{1}{\alpha} < \frac{1}{2}$ where even full ergodicity may occur. This division takes place both for the CTWF and CTWW.

## *Mean-Square Displacement and Autocorrelation Function in a Closed Form*

By applying the well-known Mellin transform and residue techniques (as shown, for example, in our earlier works [60, 71]) one can find the mean-square displacement (MSD) of the process for the long-time limit. It is a sum of regular ("*reg*") and singular ("*sing*") components,

$$\langle X(t)^2\rangle_{st}^{W} = \langle X(t)^2\rangle_{st}^{Wreg} + \langle X(t)^2\rangle_{st}^{Wsing}. \tag{23}$$

The regular (linear or Brownian) component,

$$\langle X(t)^2\rangle_{st}^{Wreg} = 2D_{st}^{W} t, \tag{24}$$

always linearly depends on time $t$, with the Brownian diffusion coefficient

$$D_{st}^{W} = \frac{1}{2}\frac{\langle x^2\rangle^{W}}{\langle \Delta t\rangle}, \tag{25}$$

where $\langle x^2\rangle^W = \langle x^2\rangle$ is given by (80) in Appendix "Stationary Case for $n = 0, 1, 2, \ldots$ and $m = 0, 1, 2, \ldots$" with $g(n = 1, m = 0) = 2$ given by (81) in Appendix "Stationary Case for $n = 0, 1, 2, \ldots$ and $m = 0, 1, 2, \ldots$" for the CTWW. The singular component is responsible for autocorrelation,

$$\langle X(t)^2\rangle_{st}^{Wsing} = \frac{2D_{st}^{Wf}}{\Gamma(\eta_1 + 1)}\left(\frac{t}{\tau_0}\right)^{\eta_1}, \tag{26}$$

where $\eta_1 \neq 1$ and the fractional diffusion coefficient

$$D_{st}^{Wf} = b_0^2\frac{\tau_0}{\langle \Delta t\rangle}\frac{1-\frac{1}{N}}{\ln N}\frac{\pi\alpha}{|\sin(\pi(\eta_1 - 1))|} \tag{27}$$

with the fractional diffusion exponent

$$\eta_1 = 1 + \alpha\left(\frac{2}{\beta} - 1\right). \tag{28}$$

The competition between both components in (23) leads to different phases depending on which component dominates at the long-time limit. Let us also take note that the horizontal red line in Fig. 4.2 delimiting the phases ED1 and ED2 from SRD1 and SRD2 corresponds to the situation $\eta_1 = 1$, which is equivalent to $\frac{1}{\beta} = \frac{1}{2}$.

The only second component in (23) is taken into account whenever we calculate velocity autocorrelation function. From (23), we get this function in the form,

$$C^W(t) = \frac{1}{2}\frac{\mathrm{d}^2}{\mathrm{d}t^2}\langle X(t)^2\rangle_{st}^W = \frac{1}{2}\frac{\mathrm{d}^2}{\mathrm{d}t^2}\langle X(t)^2\rangle_{st}^{Wsing}$$
$$= \frac{1}{\tau_0^2}\frac{\eta_1 - 1}{\Gamma(\eta_1)} D_{st}^{Wf} \frac{1}{\left(\frac{t}{\tau_0}\right)^{2-\eta_1}}, \tag{29}$$

where $\eta_1 < 2$. Keep in mind that the autocorrelation function changes from positive to negative when exponent $\eta_1$ becomes less than 1, because in our considerations coefficient $D_{st}^{Wf}$ is positive. Since both the CTWW and CTFW singular components are always present, the autocorrelation function is also current, decreasing for long-time limit according to power law. As you can see, only the singular component creates the autocorrelation function with its singular behaviour.

## *Classification of the CTRW Diffusion Phases*

In Fig. 4.3, we presented the diffusion phase diagram based on the second and fourth moments. It is a phase diagram built on the surface determined by the partial exponents $\frac{1}{\alpha}$ and $\frac{1}{\beta}$, which govern the behaviour of random walk in time and space, respectively. It contains six phases separated by straight lines (note that only four phases are marked in Fig. 4.1 in our earlier work [41]). We discuss them in succession, going from the top of the phase diagram.

(i) The Lévy diffusion (LD) phase extends above the (red) sloped straight line defined by equation $\frac{1}{\beta} = \frac{1}{2} + \frac{1}{2}\frac{1}{\alpha}$. This phase is characterized by divergent MSD for finite times and in the limit of a long time. In other words, the total mean-square displacement $\langle X(t)^2\rangle_{st}^W = \infty$ for $t > 0$.

(ii) The enhanced diffusion (ED) phase is defined by diffusion exponent $1 < \eta_1 < 2$. That is, this phase is ruled by the singular part $\langle X(t)^2\rangle_{st}^{Wsing}$, which dominates the regular part $\langle X(t)^2\rangle_{st}^{Wreg}$ for long times (i.e. $t \gg \tau_0$). This phase, extending between those mentioned above (red) sloped straight line and the (red) horizontal straight line $\frac{1}{\beta} = \frac{1}{2}$, is divided into two parts marked by ED1 and ED2. These parts are separated by the segment of the oblique straight line (blue line) given by equation $\frac{1}{\beta} = \frac{1}{4} + \frac{3}{4}\frac{1}{\alpha}$, which ends at the intersection with the horizontal red line. This equation was obtained based on the analysis of the fourth moment $\langle X(t)^4\rangle_{st}$—we return to this point in section "Discussion". As one can see, three phases extend above the horizontal line $\frac{1}{\beta} = \frac{1}{2}$ and three phases below it.

(iii) The semi-regular diffusion (SRD) phase extends between this horizontal line and the oblique one (bottom black line) defined by $\frac{1}{\beta} = \frac{1}{4} + \frac{1}{4}\frac{1}{\alpha}$. For this phase, $\langle X(t)^2\rangle_{st}^{Wreg}$ dominates $\langle X(t)^2\rangle_{st}^{Wsing}$ for long times. In other words, this phase characterizes by the singular diffusion exponent $\eta_1 < 1$ and the Brownian MSD, i.e. a linear dependence of $\langle X(t)^2\rangle_{st}^W$ on time $t$ for long times. As we argue in section "Excess Kurtosis" this phase is, however, the non-Gaussian one. It is one of the most interesting results of this work. This phase we divided into parts

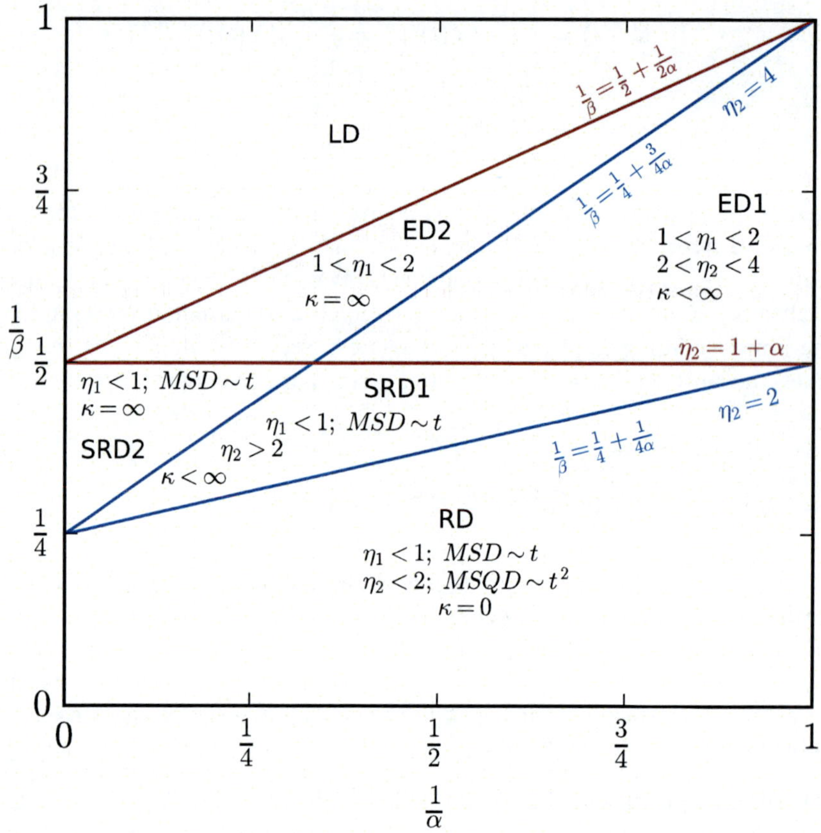


**Fig. 4.3** We present the diffusion phase diagram of the fourth order for the stationary CTWW. We show the division of the diagram into six significantly different diffusion phases. The detailed description of the diagram we posted in section "Classification of the CTRW Diffusion Phases". We remind that we use the notations: the mean-square displacement MSD $= \langle X^2 \rangle_{st}$ and the mean-square quadratic displacement MSQD $= \langle X^4 \rangle_{st}$

designated by SRD1 and SRD2 through the segment of the inclined straight line given by the equation $\frac{1}{\beta} = \frac{1}{4} + \frac{3}{4}\frac{1}{\alpha}$ and ending at the intersection with the horizontal red line. Further, the four order characteristic of this phase we consider in section "Excess Kurtosis".

(iv) Finally, the regular diffusion (RD) phase extends below the inclined straight line (bottom black line) defined by $\frac{1}{\beta} = \frac{1}{4} + \frac{1}{4}\frac{1}{\alpha}$. This phase we define by the Brownian MSD and the singular diffusion exponent $\eta_1 < 1$, similarly to the SRD1 and SRD2 phases. However, in the RD phase, kurtosis vanishes (we discuss this matter in section "Excess Kurtosis") but the other cumulants do not have to disappear. Therefore, in general, this phase should also be treated as (the higher order) non-Gaussian.

To sum up this part, we emphasize that all above-presented phases are (in general) of the non-Gaussian type and they are of the not-fBm. Besides, only the phase transition between LD and ED phases is discontinuous. Other phase transitions are continuous.

As one can see, each of the phases (except the LD) we characterize by velocity autocorrelation function relaxing according to the power law given by (29) with non-negative exponent $2 - \eta_1$.

It is worth to mention that the phase diagram of the stationary CTWW shown in Fig. 4.3 is a small part of the corresponding diffusion phase diagram of the non-stationary CTWW shown in Fig. 4.2 in our earlier work [60]. It is mainly due to the limitation in the stationary CTWW only to $\alpha > 1$. The opposite, non-stationary case $\alpha < 1$ provides increased fluctuations of time intervals playing a vital role in the spread of the phase diagram yet. These increased fluctuations allow to build a phase diagram including not an only ballistic random walk, but even Richardson's turbulent random walk of passive scalars.

### *Excess Kurtosis*

We set now the fourth moment or mean-square quadratic displacement (MSQD), $\langle X(t)^4\rangle_{st}$, in the same way as we did it for the second moment or MSD, $\langle X(t)^2\rangle_{st}$, using (22). Again, by applying the well-known Mellin transform and residue techniques (as shown, for example, in our earlier works [60, 71]), we find

$$\langle X(t)^4\rangle_{st}^{W} = \langle X(t)^4\rangle_{st}^{Wreg} + \langle X(t)^4\rangle_{st}^{Wsing}, \tag{30}$$

for the long-time limit, where $Wreg$ indexes regular and $Wsing$ the singular components. We found these components in the following forms:

$$\langle X(t)^4\rangle_{st}^{Wreg} = 12 D_{st}^{W2} t^2 \tag{31}$$

and

$$\langle X(t)^4\rangle_{st}^{Wsing} = \frac{12 D_{st}^{Wf2}}{\Gamma(\eta_2 + 1)} \left(\frac{t}{\tau_0}\right)^{\eta_2}, \tag{32}$$

where

$$D_{st}^{W2} = \left(D_{st}^{W}\right)^2, \; \eta_2 = 1 + \alpha\left(\frac{4}{\beta} - 1\right), \tag{33}$$

and the fractional super-Burnett coefficient

$$D_{st}^{Wf2} = b_0^4 \frac{\tau_0}{\langle \Delta t\rangle} \frac{1 - \frac{1}{N}}{\ln N} \frac{\pi\alpha}{\sin\left(\pi(\eta_2 - 1)\right)} \Pi(\eta_2), \tag{34}$$

where factor $\Pi(\eta_2) = (\eta_2 - 2)(\eta_2 - 3)$. This coefficient is positive within assumed range of $\eta_2 < 4$ (see Fig. 4.3 for details).

It is necessary to accept constraints imposed by the behaviour of the fourth moment or kurtosis (blue straight sloped lines). It is because we aim to deeper characterize diffusion phases of the second order (i.e. for $m = 1$), that is, driven by the finite MSD shown in Fig. 4.3 by red straight lines. It leads to six phases and not three, as would be the case for the second-order diffusion phases.

We are now considering the properties of these additional phases by which the phase diagram in Fig. 4.3 has been enriched. More specifically, the fourth moment divides the secondary phases into parts—we are considering these parts going sequentially from the top of the phase diagram.

(i) Phases located above the blue sloped line given by the equation $\frac{1}{\beta} = \frac{1}{4} + \frac{3}{4}\frac{1}{\alpha}$ are characterized by $\langle X(t)^4 \rangle = \infty$ or diverging excess kurtosis, i.e. $\kappa = \infty$. Above this line there are two phases: one located above the red horizontal line given by equation $\frac{1}{\beta} = \frac{1}{2}$, which from the top is limited by a sloping red line given by the equation: $\frac{1}{\beta} = \frac{1}{2} + \frac{1}{2}\frac{1}{\alpha}$, and the second phase below this horizontal line.

(ii) In the area of the acute angle between the blue sloped lines defined by the equations: the upper by the above mentioned in item (i) and lower utilizing $\frac{1}{\beta} = \frac{1}{4} + \frac{1}{4}\frac{1}{\alpha}$. Also, in this area of the angle, there are two different phases. The phase above the horizontal long red line (mentioned above in item (i)) characterized by diffusion exponent $\eta_1 > 1$ and the phase below it for which $\langle X(t)^2 \rangle \propto t$.

(iii) The phase lying below (mentioned above) lower diagonal blue line is characterized by $\langle X(t)^2 \rangle \propto t,\ \langle X(t)^4 \rangle \propto t^2,\ \kappa = 0,\ \eta_1 < 1$, and $\eta_2 < 2$.

Further subdivisions of the phase diagram (shown in Fig. 4.3) are discussed in section "Further Extension of the CTWW Diffusion Phase Diagram".

## *Comparison with Stationary Continuous-Time Weierstrass Flight*

For the stationary CTWW, the 1/4 of the phase diagram surface occupies the LD phase. The finite phases occupy the remaining 3/4 of the surface, i.e. the area for which MSD is limited for restricted times. Figure 4.4 shows that for the stationary CTWF the reverse situation takes place, i.e. the LD phase occupies 3/4 of the phase diagram area, while the finite phases fill the remaining 1/4 of the surface. It results in consequences that we consider below.

Technically, the analysis of the stationary CTWF is analogous to that of the stationary CTWW that's why we present our considerations only in a sketchy way. As a result, we receive equations formally identical to (23)–(29), however, solely for regions where moments for the CTWW and CTWF are simultaneously finite.

That is,

$$\langle X(t)^2\rangle_{st}^{F} = \langle X(t)^2\rangle_{st}^{Freg} + \langle X(t)^2\rangle_{st}^{Fsing}, \tag{35}$$

where

$$\langle X(t)^2\rangle_{st}^{Freg} = \langle X(t)^2\rangle_{st}^{Wreg} \tag{36}$$

and

$$\langle X(t)^2\rangle_{st}^{Fsing} = \langle X(t)^2\rangle_{st}^{Wsing} \tag{37}$$

as well

$$C^F(t) = C^W(t). \tag{38}$$

However, for the stationary CTWF, the regular component always dominates over the singular one. The latter component is still present, but it does not play a role for long times. The singular component yet, of course, lies at the core of the autocorrelation function of the process velocity.

We consider, for example, the phase placed below the sloped straight line defined by equation $\frac{1}{\beta} = \frac{1}{2m} - \frac{1}{2m}\frac{1}{\alpha}$, where $m = 1, 2, \ldots$ (cf. Fig. 4.4). The number $2m$ defines, herein, the order of the phase, where above this line or threshold the moment of the $(2m)$th order diverges but below it converges. In this sense, we can say that we deal with discontinuous phase transitions. As you can see, the phases sequentially locate or ordered descending: the higher order phase contains within the lower order phase. We are dealing with a phase diagram of $2m$ order (i.e. containing $m + 1$ phases) if we no longer divide the phase of the order of $2m$ into higher order phases. Thus, we get a tool that allows you to distinguish phases of any orders. Let's also notice that the areas of finite diffusion phases (i.e. the phases placed below the borderline defined by equation $\frac{1}{\beta} = \frac{1}{2} - \frac{1}{2}\frac{1}{\alpha}$) are now strongly reduced. That is, the stationary CTWF has a diffusion phase diagram, which is a small part (i.e. one-third) of the CTWW phase diagram (see Figs. 4.3 and 4.4 for the comparison).

Let us consider, for example, the phase of the fourth order. There are only three diffusion phases shown in Fig. 4.4. The one above the sloped straight line given by the equation $\frac{1}{\beta} = \frac{1}{2} - \frac{1}{2}\frac{1}{\alpha}$ is characterized by divergent MSD. The second phase placed below this line is the SRD2 phase, for which MSD depends on a Brownian (linear) manner from time for long times, while having kurtosis diverging. The third phase placed below the line $\frac{1}{\beta} = \frac{1}{4} - \frac{1}{4}\frac{1}{\alpha}$ is characterized by vanishing kurtosis. However, the higher cumulants in this phase exist. That's why this phase, although it is Brownian, is a non-Gaussian one.

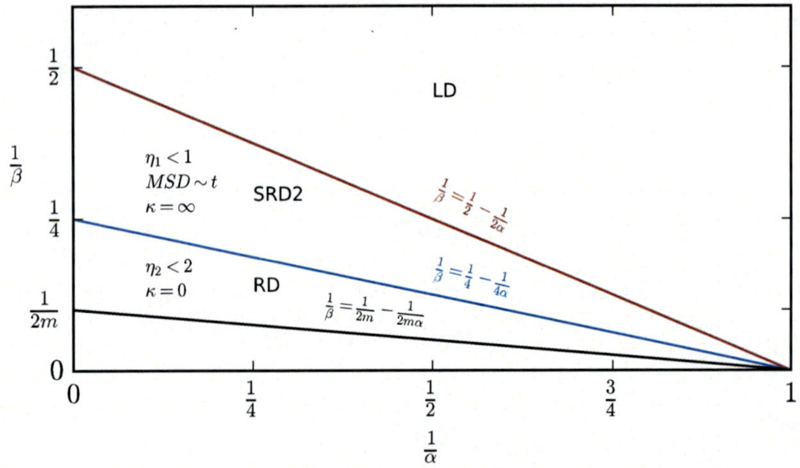


**Fig. 4.4** The diffusion phase diagram for the stationary CTWF. The plot also has a borderline $\frac{1}{\beta} = \frac{1}{2m} - \frac{1}{2m}\frac{1}{\alpha}$ for any $m$ value that allows you to build a phase diagram of the order of $2m$. Analogous boundary lines for CTWW are shown in Fig. 4.5. We divided the diagram by the borderlines of this type for $m = 1$ and $m = 2$ into significantly different diffusion phases—the detailed description of the diagram we give in the main text

## Further Extension of the CTWW Diffusion Phase Diagram

We continue to extend the results shown above for the CTWW to arbitrary even moments $\langle X(t)^{2n}\rangle$. This extension is technically more complicated than the corresponding extension for the CTWF discussed in section "Comparison with Stationary Continuous-Time Weierstrass Flight", due to the fluctuations mentioned above.

To determine any even moment of the process $X(t)$ for both CTWW and CTWF for long times in a closed form, we use the Mellin transform technique (which we outline in Appendix "Auxiliary Calculations"). This approach allows to obtain finite sum values in expressions (66) and (67) in Appendix "Auxiliary Calculations" provided that inequality occurs,

$$\frac{1}{\beta} < \frac{1}{2m} + \frac{c-2}{2m}\frac{1}{\alpha}, \quad m = 1, 2, \ldots, \tag{39}$$

where $c = 2m + 1$ for the CTWW and $c = 1$ for the CTWF.

We can check that inequality (39) provides the most reliable condition. That is, the straight boundary line defined by equality $\frac{1}{\beta} = \frac{1}{2m} + \frac{c-2}{2m}\frac{1}{\alpha}$ is the lowest on the phase diagram compared to the analogous boundary straight lines provided by the conditions imposed by the other components of MSD.

Moreover, the conjecture is assumed that the exponent governing the singular component $\langle X(t)^{2m}\rangle_{st}^{Wsing}$, takes the generic form

$$\eta_m = 1 + \alpha\left(\frac{2m}{\beta} - 1\right),\ m = 1, 2, \ldots, \tag{40}$$

where we assumed the separation

$$\langle X(t)^{2m}\rangle_{st} = \langle X(t)^{2m}\rangle_{st}^{Wreg} + \langle X(t)^{2m}\rangle_{st}^{Wsing}. \tag{41}$$

Of course, the formula (40) is a generalization of those previously derived for $m = 1$ (26) and $m = 2$ (32). Analogously, expression (41) is a generalization of (23) and (30), respectively.

We analyse the consequences of the formula (40).

(i) The area of the phase diagram defined by inequality $\eta_m > m$ refers to the case when the singular component in the formula (41) dominates over the regular component. This inequality is equivalent to the following one, $\frac{1}{\beta} > \frac{1}{2m} + \frac{m-1}{2m}\frac{1}{\alpha}$.

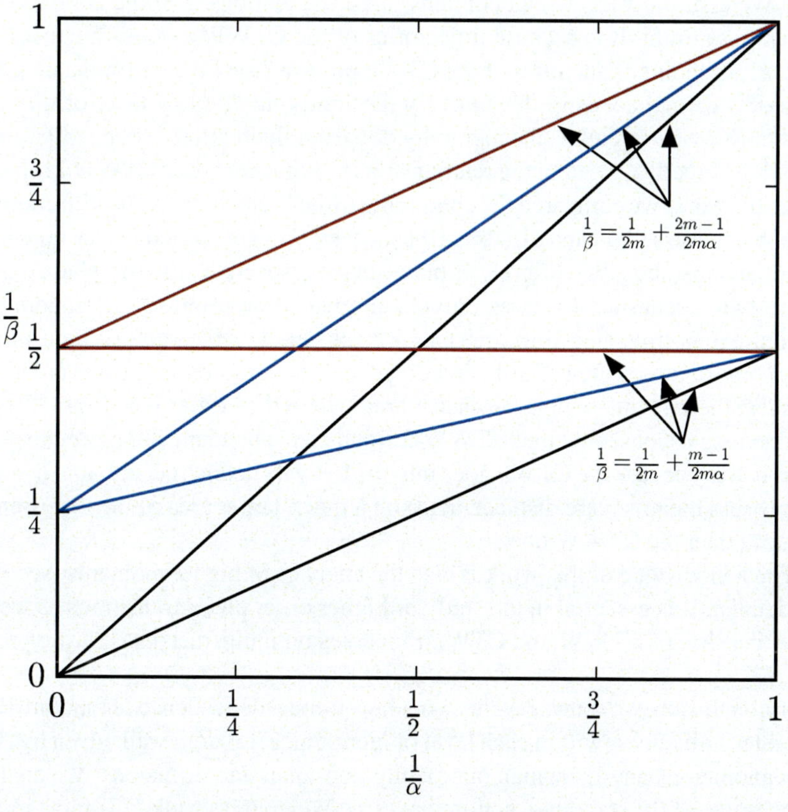


**Fig. 4.5** The diffusion phase diagram of the $2m$th order for the stationary CTWW. The arms of the acute angle with the apex of the angle at the point $\left(0, \frac{1}{2m}\right)$ determine the dominant domain of the singular component $\langle X(t)^{2m}\rangle_{st}^{Wsing}$ of the order of $2m$. The arms of the angle are described by equations, respectively: the upper by $\frac{1}{\beta} = \frac{1}{2m} + \frac{2m-1}{2m}\frac{1}{\alpha}$ and lower one by $\frac{1}{\beta} = \frac{1}{2m} + \frac{m-1}{2m}\frac{1}{\alpha}$

Hence, equality $\frac{1}{\beta} = \frac{1}{2m} + \frac{m-1}{2m}\frac{1}{\alpha}$ defines the border line below which is the area of the phase diagram (marked by SDR1 and SDR2) where the regular component in (41) dominates the singular component. In Fig. 4.3, this equation has solutions represented by the bottom red and bottom blue lines for $m = 1$ and $m = 2$, respectively. In Fig. 4.5, we have additionally presented a solution for $m = \infty$ (bottom black line).

(ii) The acute angles bounded by arms that are given above in the item (i) by the equality and (earlier introduced) equation $\frac{1}{\beta} = \frac{1}{2m} + \frac{2m-1}{2m}\frac{1}{\alpha}$ define the corresponding $2m$th order diffusion phases for the stationary CTWW.

## Concluding Remarks

This work is an extension of our previous one [41], especially when it comes to in-depth analysis of diffusion phases. The main focus of this work is the continuous-time Weierstrass walk, which contains more fluctuations than the continuous-time Weierstrass flight. It is because trajectories of the CTWW process fluctuate around the corresponding trajectories of the CTWF process (see Fig. 4.1 for details). As you can see, fluctuations can refer to every turning point at every level of the CTWW hierarchy—we can call it 'the hierarchy of partial fluctuations'. Each of partial fluctuation is directly related to the relationship (52) in Appendix "CTWW Propagator".

In this work, we compared the continuous-time Weierstrass walk with continuous-time Weierstrass flight and discussed the most significant consequences flowing here. It is especially about the differences between the respective diffusive phase diagrams. These two stochastic processes provide alternative descriptions of random motion in continuous time (see Figs. 4.1 and 4.2 for details). We showed both processes have some common area of diffusion diagram, where these formalisms are identical. Besides, there is much more extensive area where they differ (see Figs. 4.3 and 4.4 for details). We observed that CTWW formalism is abundant and more flexible than CTWF. It is because the CTWF goes into the Lévy diffusion (i.e. into the divergent of the process mean-square displacement) on a much larger area of the diffusion phase diagram than the CTWW one.

The added value of this work is also the analysis of higher moments (see Fig. 4.5 for details). It is essential in the study of higher order phase transitions as well as in the expansion of CTWW and CTWF processes on multifractal stochastic processes [72, 73].

In conclusion, we showed in this work how mutual dependence of random dynamic variables influences (within each level of hierarchical random walk given by (52)) on both anomalous and Brownian but mainly non-Gaussian diffusions. We analysed it in the frame of the stationary continuous-time Weierstrass walk. The most important consequence of this influence is autocorrelation functions' relaxations according to power laws. They are controlled by the singular exponent $\eta_1$ (see (26), (29) and (38)). The final long-term behaviour of the system is the result of competition between both worlds: regular and singular. This type of approach allows to model real-world time series for different time horizons. For example, from financial markets, where daily time series consist of alternating stationary and non-stationary fragments—the latter is defined, for example, by growing and bursting stock bubbles. Our approach can be extended to the non-stationary CTWW process—the first step in this direction we made at works [60, 74].

## Appendix 1: Expansions of Propagator

We consider expansions of the Fourier transform (or characteristic function) of the propagator,

$$\tilde{P}_{st}(k,t) = \int_{-\infty}^{\infty} \mathrm{d}X \exp(ikX) P_{st}(X,t) \tag{42}$$

for three cases.

(i) For the phases where the fourth moment diverges (doesn't exist), we have from (42), the expansion

$$\begin{aligned}\tilde{P}_{st}(k,t) &= 1 - \frac{1}{2}k^2 \int_{-\infty}^{\infty} \mathrm{d}X X^2 P_{st}(X,t) \\ &+ \int_{-\infty}^{\infty} \mathrm{d}X \sum_{n=2}^{\infty} \frac{(-1)^n}{(2n)!} k^{2n} X^{2n} P_{st}(X,t).\end{aligned} \tag{43}$$

(ii) For the phases where the second and fourth moments exist, we derive from (42), the expansion

$$\begin{aligned}\tilde{P}_{st}(k,t) &= 1 - \frac{1}{2}k^2 \int_{-\infty}^{\infty} \mathrm{d}X X^2 P_{st}(X,t) \\ &+ \frac{1}{4!}k^4 \int_{-\infty}^{\infty} \mathrm{d}X X^4 P_{st}(X,t) \\ &+ \int_{-\infty}^{\infty} \mathrm{d}X \sum_{n=3}^{\infty} \frac{(-1)^n}{(2n)!} k^{2n} X^{2n} P_{st}(X,t).\end{aligned} \tag{44}$$

(iii) In general, when the moment of the $2m$th order exists, we receive

$$\tilde{P}_{st}(k,t) = \sum_{j=0}^{m} \frac{(-1)^j}{(2j)!} k^{2j} \int_{-\infty}^{\infty} \mathrm{d}X X^{2j} P_{st}(X,t) + \tilde{R}_{st}^{2(m+1)}(k,t), \tag{45}$$

where

$$\tilde{R}_{st}^{2(m+1)}(k,t) = = \int_{-\infty}^{\infty} \mathrm{d}X \sum_{n=m+1}^{\infty} \frac{(-1)^n}{(2n)!} k^{2n} X^{2n} P_{st}(X,t), \tag{46}$$

we remind that all odd moments vanish.

In all above formulas (43), (44) and (45) it is not possible to change the integration with summation in last components and to integrate the term after the term one by one, because the obtained integrals (moments of variable X) are divergent for $n > m$. Moreover, the integral of the sum of the odd terms disappeared, because the propagator $P(X,t)$ is an even function of the variable $X$ (no drift is present in both random walks).

As for the notations, the second ingredient in (43) and (44) is the second moment $\langle X(t)^2 \rangle$, the third ingredient in (44) is the fourth moment $\langle X(t)^4 \rangle$, while the last ingredient in both equations is the rest $\tilde{R}_{st}^{2(n+1)}(k,t)$ of the expansions, where $n = 1$ for (43), $n = 2$ for (44) and, in general, $n = m + 1$ for (45). Indeed, this notation is used in section "Propagator".

You also need to pay attention to the important property of the rest defined by the second component in (45),

$$\frac{\mathrm{d}^{2m+1}}{\mathrm{d}k^{2m+1}} \tilde{R}_{st}^{2m+1}(k,t)|_{k=0} = 0. \tag{47}$$

It results directly from the definition of the rest, where variable $k^{2(m+1)}$ can be pulled out in front of the integral (see (43) and (44) for characteristic examples).

## *Exact Stationary CTWW and CTWF Propagators*

### CTWW Propagator

According to generic (19), specific expression for the stationary propagator $\hat{\tilde{P}}_{st}^{W}(k,u)$ of the CTWW takes the form

$$\hat{\tilde{P}}_{st}^{W}(k,u) = \hat{\tilde{\Xi}}^{W}(k,u) + \hat{\tilde{\chi}}^{W}(k,u)\hat{\tilde{P}}^{W}(k,u), \tag{48}$$

where

$$
\begin{aligned}
\hat{\tilde{\Xi}}^W(k,u) &= \tau_0\left(1-\frac{\tau}{N}\right)\sum_{j=0}^{\infty}\left(\frac{\tau^2}{N}\right)^j\hat{\tilde{\Theta}}_j^W(k,u),\\
\hat{\tilde{\chi}}^W(k,u) &= \left(1-\frac{\tau}{N}\right)\sum_{j=0}^{\infty}\left(\frac{\tau}{N}\right)^j\hat{\tilde{\Theta}}_j^W(k,u),\\
\hat{\tilde{P}}^W(k,u) &= \frac{\hat{\tilde{\Psi}}^W(k,u)}{1-\hat{\tilde{\psi}}^W(k,u)}\\
&= \tau_0\frac{\sum_{j=0}^{\infty}\left(\frac{\tau}{N}\right)^j\hat{\tilde{\Theta}}_j^W(k,u)}{\sum_{j=0}^{\infty}\left(\frac{1}{N}\right)^j\left(1-\hat{\tilde{\Theta}}_j^W(k,u)\right)},
\end{aligned}
\tag{49}
$$

with

$$
\hat{\tilde{\Theta}}_j^W(k,u) = \frac{u\tau_0\tau^j+1}{(u\tau_0\tau^j+1)^2+k^2(b_0b^j)^2}
\tag{50}
$$

and

$$
\begin{aligned}
\hat{\tilde{\psi}}^W(k,u) &= \left(1-\frac{1}{N}\right)\sum_{j=0}^{\infty}\left(\frac{1}{N}\right)^j\hat{\tilde{\Theta}}_j^W(k,u),\\
\hat{\tilde{\Psi}}^W(k,u) &= \tau_0\left(1-\frac{1}{N}\right)\sum_{j=0}^{\infty}\left(\frac{\tau}{N}\right)^j\hat{\tilde{\Theta}}_j^W(k,u).
\end{aligned}
\tag{51}
$$

The form of propagator given by (48) is the result of a hierarchical spatial-temporal coupling and a simple dynamic coupling between $x$ and $\Delta t$ at each level $j$ of the hierarchy as follows:

$$
x(\Delta t) = \pm v_0 v^j \Delta t
\tag{52}
$$

resulting from $\delta$-s present in the definition (14), where $b_0 = v_0\tau_0$ and $b = v\tau$. Each sign in (52) is drawn with equal probability 1/2, because there is no drift in the system.

Let us emphasize that (52) is the equation of discrete stochastic dynamics without drift, where the single-step displacement $x(\Delta t) = X(t+\Delta t) - X(t)$, the probability of its orientation $\pm$ is $prob(+) = prob(-) = 0.5$, its random discrete variable (or index) $j$ comes from the distribution $w(j) = \left(1-\frac{1}{N}\right)\frac{1}{N^j}$ and the single-step interevent time $\Delta t$ from the exponential distribution $p_j(\Delta t) = \frac{1}{\tau_0\tau^j}\exp\left(-\Delta t/\tau_0\tau^j\right)$. As one can see, the stochasticity of this dynamics is governed by three random variables drawn from three different probability distributions. What's more, (52) is the equation of discrete stochastic dynamics with a random time step.

Notably, for the CTWF the analogous discrete stochastic dynamic equation takes a much simpler form

$$x(\Delta t) = \pm b_0 b^j, \tag{53}$$

where $b_0 = v_0\tau_0$ and $b = v\tau$.

Both stationary and non-stationary CTWWs are characterized by (i) the random nature of $\Delta t$ variable, as this variable is subject to the exponential distribution $p_j(\Delta t)$; (ii) time-space coupling (dependence) defined by (52) at any level of walk's hierarchy and (iii) the hierarchical nature of this walk. Hence, even function of $x$ is cross-correlated with any function of $\Delta t$ (see Appendix "Stationary Case for $n = 0, 1, 2, \ldots$ and $m = 0, 1, 2, \ldots$" for details). To see that variables $x$ and $\Delta t$ are uncorrelated, just write the cross-correlation function in the form $\langle x\Delta t\rangle - \langle x\rangle\langle\Delta t\rangle = \langle v_0 v^j\rangle\left[\langle(\Delta t)^2\rangle - \langle\Delta t\rangle^2\right] = 0$ as a dichotomic averaging $\langle v_0 v^j\rangle = +\frac{1}{2}v_0 v^j - \frac{1}{2}v_0 v^j = 0$ (no drift is present in the system).

The list of items mentioned above has consequences that are the subject of this work. We examine these consequences in comparison with the corresponding ones of the CTWF model, for which dependence (52) is replaced by (53). For the CTWF, arbitrary joint function of $x$ and $\Delta t$ are independent at any level $j$ of the hierarchy. For both models, arbitrary even function of $x$ is cross-correlated with any functions of $\Delta t$ (see Appendix "Non-factorial Spatial-Temporal Cross-Moments" for details).

Note that equality occurs

$$\hat{\tilde{\Theta}}{}_j^W(k=0,u) = \frac{1}{u\tau_0\tau^j + 1}, \tag{54}$$

which is useful. Hence, and from the last equality in (49), we get

$$\hat{\tilde{P}}{}^W(k=0,u) = \frac{1}{u}. \tag{55}$$

## CTWF Propagator

For the CTWF, analogical relations take place. Namely, (48) takes the form

$$\hat{\tilde{P}}{}_{st}^F(k,u) = \hat{\tilde{\Xi}}{}^F(k,u) + \hat{\tilde{\chi}}{}^F(k,u)\hat{\tilde{P}}{}^F(k,u), \tag{56}$$

where

$$\hat{\tilde{\Xi}}{}^F(k,u) = \tau_0\left(1 - \frac{\tau}{N}\right)\sum_{j=0}^{\infty}\left(\frac{\tau^2}{N}\right)^j \hat{\tilde{\Theta}}{}_j^F(k,u),$$

$$\hat{\tilde{\chi}}^F(k,u) = \left(1 - \frac{\tau}{N}\right)\sum_{j=0}^{\infty}\left(\frac{\tau}{N}\right)^j \hat{\tilde{\Theta}}_j^F(k,u),$$

$$\hat{\tilde{P}}^F(k,u) = \frac{\hat{\tilde{\Psi}}^F(k=0,u)}{1 - \hat{\tilde{\psi}}^F(k,u)}$$

$$= \tau_0 \frac{\sum_{j=0}^{\infty}\left(\frac{\tau}{N}\right)^j \hat{\tilde{\Theta}}_j^F(k=0,u)}{\sum_{j=0}^{\infty}\left(\frac{1}{N}\right)^j \left(1 - \hat{\tilde{\Theta}}_j^F(k,u)\right)}, \tag{57}$$

with

$$\hat{\tilde{\Theta}}_j^F(k,u) = \cos(kb_0 b^j)\hat{\tilde{\Theta}}_j^W(k=0,u) \tag{58}$$

and

$$\hat{\tilde{\psi}}^F(k,u) = \left(1 - \frac{1}{N}\right)\sum_{j=0}^{\infty}\left(\frac{1}{N}\right)^j \hat{\tilde{\Theta}}_j^F(k,u),$$

$$\hat{\tilde{\Psi}}^F(k,u) = \tau_0\left(1 - \frac{1}{N}\right)\sum_{j=0}^{\infty}\left(\frac{\tau}{N}\right)^j \hat{\tilde{\Theta}}_j^F(k,u), \tag{59}$$

which are simpler as (58) has simpler form than the corresponding (50).

Thanks to (58), we obtain useful properties regarding marginal distributions. Namely, we have the following relations for marginal quantities:

$$\hat{\tilde{\Theta}}_j^F(k=0,u) = \hat{\tilde{\Theta}}_j^W(k=0,u),$$
$$\hat{\tilde{\psi}}^F(k=0,u) = \hat{\tilde{\psi}}^W(k=0,u),$$
$$\hat{\tilde{\Psi}}^F(k=0,u) = \hat{\tilde{\Psi}}^W(k=0,u),$$
$$\hat{\tilde{P}}^F(k=0,u) = \hat{\tilde{P}}^W(k=0,u),$$
$$\hat{\tilde{\Xi}}^F(k=0,u) = \hat{\tilde{\Xi}}^W(k=0,u),$$
$$\hat{\tilde{\chi}}^F(k=0,u) = \hat{\tilde{\chi}}^W(k=0,u),$$
$$\hat{\tilde{P}}_{st}^F(k=0,u) = \hat{\tilde{P}}_{st}^W(k=0,u),$$
$$\frac{\mathrm{d}^{2m+1}}{\mathrm{d}k^{2m+1}}\hat{\tilde{\Theta}}_j^F(k,u)\,|_{k=0} = \frac{\mathrm{d}^{2m+1}}{\mathrm{d}k^{2m+1}}\hat{\tilde{\Theta}}_j^W(k,u)\,|_{k=0} = 0,$$
$$\frac{\mathrm{d}^{2m}}{\mathrm{d}k^{2m}}\hat{\tilde{\Theta}}_j^W(k,u)\,|_{k=0} = (-1)^m 2m!(b_0)^{2m} b^{2mj} \times \left[\hat{\tilde{\Theta}}_j^W(k=0,u)\right]^{2m+1},$$
$$\frac{\mathrm{d}^{2m}}{\mathrm{d}k^{2m}}\hat{\tilde{\Theta}}_j^F(k,u)\,|_{k=0} = (-1)^m (b_0)^m b^{2mj}\hat{\tilde{\Theta}}_j^F(k=0,u). \tag{60}$$

## Auxiliary Calculations

We carry out calculations that form the basis for the closed expression containing an infinite summation after the current variable $j$. For example, a typical case is

$$\begin{aligned}
&\frac{d^{2m}}{dk^{2m}}\hat{\tilde{\Xi}}^{W}(k,u)\,|_{k=0} \\
&= \tau_0\left(1-\frac{\tau}{N}\right)\sum_{j=0}^{\infty}\left(\frac{\tau^2}{N}\right)^j \frac{d^{2m}}{dk^{2m}}\hat{\tilde{\Theta}}^{W}_j(k,u)\,|_{k=0} \\
&= (-1)^m 2m!(b_0)^{2m}\tau_0\left(1-\frac{\tau}{N}\right)\sum_{j=0}^{\infty}\left(\frac{\tau^2 b^{2m}}{N}\right)^j \\
&\times\left(\frac{1}{u\tau_0\tau^j+1}\right)^{2m+1},\quad m=1,2,\ldots. \qquad (61)
\end{aligned}$$

To perform summation, we replace $\left(\frac{1}{u\tau_0\tau^j+1}\right)^{2m+1}$ by its inverse Mellin transformation so that instead of (61), we can write

$$\begin{aligned}
&\frac{d^{2m}}{dk^{2m}}\hat{\tilde{\Xi}}^{W}(k,u)\,|_{k=0} \\
&= (-1)^m 2m!(b_0)^{2m}\tau_0\left(1-\frac{\tau}{N}\right)\sum_{j=0}^{\infty}\left(\frac{\tau^2 b^{2m}}{N}\right)^j \\
&\times\frac{1}{2\pi i}\int_{c-i\infty}^{c+i\infty} ds\,\tau^{-sj}\frac{\pi}{\sin(\pi s)}(u\tau_0)^{-s}\binom{s-1}{2m}, \qquad (62)
\end{aligned}$$

where $0 < c = \Re\, s < 2m+1$ and $\binom{s-1}{2m}$ is the generalized binomial (Newton) coefficient.

The sum and integral can be interchanged in (62) only when they converge. After such a change, we are dealing with a geometric series, $\left(\frac{b^{2m}}{\tau^{s-2}N}\right)^j$, which is convergent if and only if its quotient $\frac{b^{2m}}{\tau^{c-2}N} < 1$. This inequality is equivalent (to good approximation) to the following essential one:

$$\frac{1}{\beta} < \frac{1}{2m} + \frac{2m-1}{2m}\frac{1}{\alpha}, \qquad (63)$$

where the upper limit $c = 2m+1$ was used.

Thus, if constrain (63) is satisfied then (62) takes the form

$$\frac{\mathrm{d}^{2m}}{\mathrm{d}k^{2m}}\hat{\hat{\Xi}}^{W}(k,u)\mid_{k=0}=(-1)^{m}(2m)!(b_0)^{2m}\tau_0\left(1-\frac{\tau}{N}\right)$$
$$\times\frac{1}{2\pi i}\int_{c-i\infty}^{c+i\infty}\mathrm{d}s\frac{1}{1-\frac{b^{2m}}{\tau^{s-2}N}}\frac{\pi}{\sin(\pi s)}\binom{s-1}{2m}(u\tau_0)^{-s}. \tag{64}$$

This form exhibits the origin of singularities coming from the hierarchical structure of the process random walk which is clustered according to the geometric series. The integrand has poles of the first order from $\frac{1}{\sin(\pi s)}$ at $s=s_0=0,-1,-2,\ldots,$ and from $\frac{1}{1-\frac{b^{2m}}{\tau^{s-2}N}}$ at $s=s_1(n)=2+\alpha\left(\frac{2m}{\beta}-1\right)\pm 2\pi i\frac{n}{\ln\tau}$, $n=0,1,\ldots$ Integration is performed using the well-known residual method, as it is possible to build a contour (here rectangular) encompassing these poles. The poles $s_0=0$ and $s_1(n=0)$ are still selected, because the remaining ones give a vanishing contribution to the integral (64) for $\mid u\mid\ll 1$, which we show below. The right side of the contour is parallel to the vertical imaginary axis and has a real coordinate just equal to $c$. It shows that the integral on the other sides of the contour disappear as they move away to infinity. Hence, (64) takes the form

$$\frac{\mathrm{d}^{2m}}{\mathrm{d}k^{2m}}\hat{\hat{\Xi}}^{W}(k,u)\mid_{k=0}=(-1)^{m}(2m)!(b_0)^{2m}\tau_0\left(1-\frac{\tau}{N}\right)$$
$$\times\left[\frac{1}{1-\frac{\tau^2b^{2m}}{N}}+\frac{1}{\ln N}\frac{\pi\alpha}{\sin(\pi s_1)}\binom{s_1-1}{2m}(u\tau_0)^{-s_1}\right], \tag{65}$$

where simplified notation $s_1(n=0)=s_1$ was used.

## *The Key Marginal Equalities*

We present a source of key equations that define the upper boundaries of the MSD $<\infty$ areas on CTWW and CTWF phase diagrams. It can be shown that these upper boundaries are dictated by the behaviour of $\hat{\hat{\Xi}}(k,u)$. The appropriate derivatives of $\hat{\hat{\Xi}}(k,u)$ we present in the definition of any even moments of the process $X(t)$ (cf. (19) and (21) and (22) for example). For the CTWW, we have

$$\frac{\mathrm{d}^{2m}}{\mathrm{d}k^{2m}}\hat{\hat{\Xi}}^{W}(k,u)\mid_{k=0}=(-1)^{m}2m!(b_0)^{2m}\tau_0\left(1-\frac{\tau}{N}\right)$$
$$\times\sum_{j=0}^{\infty}\left(\frac{\tau^2b^{2m}}{N}\right)^{j}\left[\hat{\hat{\Theta}}_{j}^{W}(k=0,u)\right]^{2m+1}, \tag{66}$$

while for the CTWF, we get simpler form

$$\frac{\mathrm{d}^{2m}}{\mathrm{d}k^{2m}}\hat{\tilde{\Xi}}^F(k,u)\mid_{k=0}=\tau_0\left(1-\frac{\tau}{N}\right)$$
$$\times\sum_{j=0}^{\infty}\left(\frac{\tau^2 b^{2m}}{N}\right)^j\hat{\tilde{\Theta}}_j^F(k=0,u). \tag{67}$$

The fact that $\hat{\tilde{\Theta}}_j^F(k=0,u)$ for CTWF occurs in the first power independently of $m$ makes difference between CTWF and CTWW essentials. The existence of $\hat{\tilde{\Theta}}_j^W(k=0,u)$ for CTWW in power of $m$ is the result of a simple spatial-temporal coupling (52) at each level $j$ of the random walk hierarchy. For the CTWF such a coupling is absent.

## Appendix 2: Basic Distributions and Means

### *Basic Distributions*

The waiting-time and displacement distribution $\psi(x,\Delta t)$ is the primary distribution. The non-stationary and stationary CTRW formalisms (including CTWW and CTWF ones, which are typical hierarchical cases) we based on it. It is the probability density of the walker single-step displacement $x$ in its duration time $\Delta t$. After this time, the walker begins its next single step by marking the turning point of its trajectory. In both our cases (defined by walks and flights), we represent corresponding WTDDs in the form of the Weierstrass distributions.

Our approach is two steps: first, we define below WTDDs for the non-stationary CTWF and CTWW in the forms,

$$\psi(x,\Delta t)=\frac{1}{2}\left(1-\frac{1}{N}\right)\sum_{j=0}^{\infty}\frac{1}{N^j}\left[\delta(x-z_j)-\delta(x+z_j)\right]$$
$$\times\frac{1}{\tau_0\tau^j}\exp\left(-\frac{\Delta t}{\tau_0\tau^j}\right)$$
$$=\begin{cases}\psi^F(x,\Delta t), & \text{where } z_j=b_0 b^j \text{ for CTWF},\\ \psi^W(x,\Delta t), & \text{where } z_j=v_0 v^j\Delta t \text{ for CTWW},\end{cases} \tag{68}$$

respectively.

The initial step of the walker requires special treatment for the stationary situations—this is our second step. That is, a proper averaging over initial time of the process is required as the time origin can be chosen arbitrarily in this situation. Applying the conditional probability techniques [14] (see also [75]), one can define auxiliary quantities,

$$\chi(\Delta t) = \int_{-\infty}^{\infty} \mathrm{d}x \chi(x, \Delta t) = \frac{\Psi(\Delta t)}{\langle \Delta t \rangle} \tag{69}$$

and

$$\begin{aligned} \Psi(\Delta t) &= \int_{-\infty}^{\infty} \mathrm{d}x \Psi(x, \Delta t) = \int_{\Delta t}^{\infty} \mathrm{d}\Delta t' \psi(\Delta t'), \\ \psi(\Delta t) &= \int_{-\infty}^{\infty} \mathrm{d}x \psi(x, \Delta t), \end{aligned} \tag{70}$$

where for the non-stationary sojourn probabilities, we get

$$\begin{aligned} \Psi(x, \Delta t) &= \frac{1}{2}\left(1 - \frac{1}{N}\right) \sum_{j=0}^{\infty} \frac{1}{N^j} \left[\delta(x - z_j) - \delta(x + z_j)\right] \exp\left(-\frac{\Delta t}{\tau_0 \tau^j}\right) \\ &= \begin{cases} \Psi^F(x, \Delta t), & \text{where } z_j = b_0 b^j \text{ for CTWF}, \\ \Psi^W(x, \Delta t), & \text{where } z_j = v_0 v^j \Delta t \text{ for CTWW}. \end{cases} \end{aligned} \tag{71}$$

From (69)–(71), we obtain the sought stationary WTDDs in the forms,

$$\begin{aligned} \chi(x, \Delta t) &= \frac{1}{2}\left(1 - \frac{\tau}{N}\right) \frac{1}{\tau_0} \sum_{j=0}^{\infty} \frac{1}{N^j} \left[\delta(x - z_j) - \delta(x + z_j)\right] \exp\left(-\frac{\Delta t}{\tau_0 \tau^j}\right) \\ &= \begin{cases} \chi^F(x, \Delta t), & \text{where } z_j = b_0 b^j \text{ for CTWF}, \\ \chi^W(x, \Delta t), & \text{where } z_j = v_0 v^j \Delta t \text{ for CTWW}. \end{cases} \end{aligned} \tag{72}$$

Analogously, from (72), we obtain the sought stationary sojourn probabilities in the forms,

$$\begin{aligned} \Xi(x, \Delta t) &= \frac{1}{2}\left(1 - \frac{\tau}{N}\right) \sum_{j=0}^{\infty} \left(\frac{\tau}{N}\right)^j \left[\delta(x - z_j) - \delta(x + z_j)\right] \exp\left(-\frac{\Delta t}{\tau_0 \tau^j}\right) \\ &= \begin{cases} \Xi^F(x, \Delta t), & \text{where } z_j = b_0 b^j \text{ for CTWF}, \\ \Xi^W(x, \Delta t), & \text{where } z_j = v_0 v^j \Delta t \text{ for CTWW}, \end{cases} \end{aligned} \tag{73}$$

because the relationship between $\chi$ and $\Xi$, both for CTWF and CTWW, is analogous to the relationship between $\psi$ and $\Psi$ (the second equality in the first row of (70)).

As one can see, WTDDs for CTWF and CTWW (both for non-stationary and stationary cases) differ in spatial parts. In the former case, the relationship between $x$ and $\Delta t$ does not occur while in the latter case, there is a simple relationship. Recall, we have on any level of $j$ a dichotomic relation $x = \pm v_0 v^j \Delta t$. The difference between both relations leads to significant differences between random walks.

### *Marginal Distributions*

From (68) and (71)–(73), we get the list of marginal, solely time-dependent distributions by integrating both sides of these equations over single-step variable $x$,

$$\psi^F(\Delta t) = \psi^W(\Delta t) = \left(1 - \frac{1}{N}\right) \sum_{j=0}^{\infty} \frac{1}{N^j} \frac{1}{\tau_0 \tau^j} \exp\left(-\frac{\Delta t}{\tau_0 \tau^j}\right), \tag{74}$$

$$\Psi^F(\Delta t) = \Psi^W(\Delta t) = \left(1 - \frac{1}{N}\right) \sum_{j=0}^{\infty} \frac{1}{N^j} \exp\left(-\frac{\Delta t}{\tau_0 \tau^j}\right), \tag{75}$$

$$\chi^F(\Delta t) = \chi^W(\Delta t) = \left(1 - \frac{\tau}{N}\right) \frac{1}{\tau_0} \sum_{j=0}^{\infty} \frac{1}{N^j} \frac{1}{\exp} \left(-\frac{\Delta t}{\tau_0 \tau^j}\right), \tag{76}$$

$$\Xi^F(\Delta t) = \Xi^W(\Delta t) = \left(1 - \frac{1}{N}\right) \sum_{j=0}^{\infty} \left(\frac{\tau}{N}\right)^j \exp\left(-\frac{\Delta t}{\tau_0 \tau^j}\right). \tag{77}$$

From (68), we get marginal, space-dependent distribution by integrating both sides of above equation over single-step variable $\Delta t$,

$$\psi^F(x) = \frac{1}{2} \left(1 - \frac{1}{N}\right) \sum_{j=0}^{\infty} \frac{1}{N^j} \frac{1}{b_0 b^j} \delta\left(\frac{|x|}{b_0 b^j} - 1\right), \tag{78}$$

$$\psi^W(x) = \frac{1}{2} \left(1 - \frac{1}{N}\right) \sum_{j=0}^{\infty} \frac{1}{N^j} \frac{1}{b_0 b^j} \exp\left(-\frac{|x|}{b_0 b^j}\right), \tag{79}$$

where $b_0 = v_0 \tau_0$ and $b = v\tau$. Analogously, we can easily get other distributions: $\chi^{F,W}(x)$, $\Psi^{F,W}(x)$ and $\Xi^{F,W}(x)$ in similar forms.

## *Non-factorial Spatial-Temporal Cross-Moments*

We mark cross-moments and hence boundary moments for CTWF and CTWW processes both for the non-stationary and stationary cases.

### Non-stationary Case for $n = 0, 1, 2, \ldots$ and $m = 0, 1, 2, \ldots$

The spatial-temporal cross-moment takes the following form in the non-stationary case:

$$\langle x^{2n}\Delta t^m\rangle = \int_0^\infty \mathrm{d}\Delta t\,\Delta t^m \int_{-\infty}^{\infty} \mathrm{d}x x^{2n}\psi(\Delta x, \Delta t)$$
$$= \begin{cases} g(n,m) b_0^{2n}\tau_0^m \frac{1-\frac{1}{N}}{1-\frac{b^{2n}\tau^m}{N}}, \text{ for } \frac{1}{\beta} < \frac{1}{2n} - \frac{m}{2n}\frac{1}{\alpha}, \\ \infty, \qquad \text{otherwise,} \end{cases} \tag{80}$$

where

$$g(n,m) = \begin{cases} m!, & \text{for CTWF process,} \\ (2n+m)!, & \text{for CTWW process.} \end{cases} \tag{81}$$

Now, we can formulate the following properties:

(i) The appearance of the factor $(2n + m)!$ instead of $m!$ it results directly from dependency (52).
(ii) Hence, the spatial boundary moment $\langle x^{2n}\rangle$ can be obtained by setting $m = 0$, while the temporal $\langle \Delta t^m\rangle$ one by setting $n = 0$.
(iii) As one can see, the cross-moment (80) is (in general) non-factorial. It means that spatial-temporal cross-correlations reside in the CTWF and CTWW processes. The basic source of these correlations is the hierarchical nature of both processes.

### Stationary Case for $n = 0, 1, 2, \ldots$ and $m = 0, 1, 2, \ldots$

For the stationary case, we have different spatial-temporal cross moments although similarly calculated,

$$\langle x^{2n}\Delta t^m\rangle_{st} = \int_0^\infty \mathrm{d}\Delta t\,\Delta t^m \int_{-\infty}^{\infty} \mathrm{d}x x^{2n}\chi(\Delta x, \Delta t)$$
$$= \begin{cases} g(n,m) b_0^{2n}\tau_0^m \frac{1-\frac{\tau}{N}}{1-\frac{b^{2n}\tau^{m+1}}{N}}, \text{ for } \frac{1}{\beta} < \frac{1}{2n} - \frac{m+1}{2n}\frac{1}{\alpha} \\ \infty, \qquad \text{otherwise,} \end{cases} \tag{82}$$

where $g(n, m)$ is given by (81).

These cross-moments have properties analogous to those (i)–(iii) presented above for the non-stationary case as there is a simple relation between both types of cross-moments in the form,

$$\begin{aligned}\langle x^{2n}\Delta t^m\rangle_{st} &= \frac{g(n,m)}{g(n,m+1)}\frac{\langle x^{2n}\Delta t^{m+1}\rangle}{\langle \Delta t\rangle}\\ &\neq \frac{g(n,m)}{g(n,m+1)}\langle x^{2n}\Delta t^m\rangle\end{aligned} \tag{83}$$

and

$$\langle x^{2n}\Delta t^m\rangle_{st} \neq \langle x^{2n}\rangle\langle \Delta t^m\rangle_{st}. \tag{84}$$

Hence, we get expressions for stationary boundary moments in the form

$$\langle \Delta t\rangle_{st} = \frac{g(0,1)}{g(0,2)}\frac{\langle \Delta t^2\rangle}{\langle \Delta t\rangle} \neq \langle \Delta t\rangle \tag{85}$$

and

$$\langle x^2\rangle_{st} = \frac{g(1,0)}{g(1,1)}\frac{\langle x^2\Delta t\rangle}{\langle \Delta t\rangle} \neq \langle x^2\rangle. \tag{86}$$

The cross-correlations (both for stationary and non-stationary cases) are non-vanishing for any even function of $x$-variable. However, for odd functions of $x$ the cross-correlations vanish, because we are dealing with random walks without drifts.

## References


1. E.W. Montroll, G.H. Weiss, J. Math. Phys. **6**(2), 167 (1965)
2. J. Klafter, G. Zumofen, J. Phys. Chem. **98**(30), 7366 (1994)
3. V. Zaburdaev, S. Denisov, J. Klafter, Rev. Modern Phys. **87**, 483 (2015)
4. P. Paradisi, G. Kaniadakis, A.M. Scarfone, Chaos Solitons, and Fractals **81**, 407 (2015)
5. H. Scher, Eur. Phys. J. B **90**, 252 (2017)
6. M.F. Schlesinger, Eur. Phys. J. B **90**, 93 (2017)
7. R. Kutner, J. Masoliver, Eur. Phys. J. B. **90**, 50 (2017)
8. H. Scher, M. Lax, J. Non-Crystalline Solids **8–10**, 497 (1972)
9. H. Scher, M. Lax, Phys. Rev. B **7**, 4491 (1973)
10. E.W. Montroll, B.J. West, in *Fluctuation Phenomena*, Studies in Statistical Mechanics, ed. by E.W. Montroll, J.L. Lebowitz (eds.), vol. VII, Chap. 2 (Elsevier (North-Holland Publishing Company), Amsterdam, 1979), p. 61
11. G.H. Weiss, R.J. Rubin, in *Advances in Chemical Physics*, vol. LII, ed. by I. Prigogine, S.A. Rice (Wiley, New York, 1983), p. 363
12. E.W. Montroll, M.F. Shlesinger, in *Nonequilibrium Phenomena II. From Stochastics to Hydrodynamics*. Studies in Statistical Mechanics, vol. XI, Chap. 1, ed. by J.L. Lebowitz, E.W. Montroll (Elsevier (North-Holland Publishing Company), Amsterdam, 1984), p. 1

13. G.H. Weiss, in *Fractals in Science*, ed. by A. Bunde, S. Havlin (Springer, Berlin, 1995), p. 118
14. J.W. Haus, K.W. Kehr, Phys. Rep. **150**(5 & 6), 141 (1987)
15. M. Montero, J. Masoliver, Phys. Rev. E **76**, 061115–1 (2007)
16. K.W. Kehr, R. Kutner, K. Binder, Phys. Rev. B **23**, 4931 (1981)
17. R. Kutner, J. Phys. C: Solid State Phys. **18**, 6323 (1985)
18. V.M. Kenkre, E.W. Montroll, M.F. Shlesinger, J. Stat. Phys. **10**(5), 421 (1974)
19. V.M. Kenkre, R.S. Knox, Phys. Rev. B **9**, 5279 (1974)
20. V.M. Kenkre, Phys. Rev. B **12**, 2150 (1975)
21. U. Landmann, E.W. Montroll, M.F. Shlesinger, Proc. Natl. Acad. Sci. USA **74**, 430 (1977)
22. U. Landmann, M.F. Shlesinger, Phys. Rev. B **19**, 6220 (1979)
23. D.T. Gillespie, Phys. Lett. A **64**, 22 (1977)
24. M.O. Caceres, H.S. Wio, Z. Phys. B **54**, 175 (1984)
25. J. Masoliver, G. Weiss, Phys. A **183**, 537 (1990)
26. J. Klafter, R. Silbey, Phys. Rev. Lett. **44**, 55 (1980)
27. J.K.E. Tunaley, Phys. Rev. Lett. **33**, 1037 (1974)
28. J.K.E. Tunaley, J. Stat. Phys. **14**, 461 (1976)
29. M.M. Meerscheart, P. Straka, Ann. Prob. **42**(4), 1699 (2014)
30. V.S. Barbu, N. Limnios, *Lecture Notes in Statistical Physics*, vol. 192 (Springer, Berlin, 2008)
31. M. Scher, E.W. Montroll, Phys. Rev. B **12**, 2455 (1975)
32. G. Pfister, H. Scher, Adv. Phys. **27**(5), 747 (1978)
33. M. Esposito, K. Lindenberg, Phys. Rev. E **77**, 051119 (2008)
34. J.-P. Bouchaud, A. Georges, Anomalous diffusion in disordered media: statistical mechanics. models and physical applications. Phys. Rep. **195**(4, 5), 127 (1990)
35. I.M. Sokolov, J. Klafter, Chaos **15**(2), 26103 (2005)
36. J. Klafter, G. Zumofen, Phys. Rev. E **49**, 4873 (1994)
37. V. Zaburdaev, K. Chukbar, JETP **94**, 252 (2002)
38. J. Klafter, I. Sokolov, *First Steps in Random Walks: From Tools to Applications* (Oxford University Press, Oxford, 2011)
39. J. Klafter, A. Blumen, M.F. Shlesinger, Phys. Rev. A **35**, 3081 (1987)
40. G. Zumofen, J. Klafter, Phys. Rev. E **47**, 851 (1993)
41. R. Kutner, F. Świtała, Quant. Fin. **3**, 201 (2003)
42. E. Barkai and J. Klafter, in *Chaos, Kinetics and Nonlinear Dynamics in Fluids and Plasmas*, ed. by S. Benkadda, G.M. Zaslavsky (Springer, Berlin, 1998), p. 373
43. V.Y. Zaburdaev, M. Schmiedeberg, H. Stark, Phys. Rev. E **78** 011119
44. S. Denisov, V. Zaburdaev, P. Hänggi, Phys. Rev. Lett. **109**, 069903(E) (2012)
45. V.Y. Zaburdaev, S. Denisov, P. Hänggi, Phys. Rev. Lett. **106**, 180601 (2011)
46. V.Y. Zaburdaev, S. Denisov, P. Hänggi, Phys. Rev. Lett. **110**, 170604 (2013)
47. P. Tong, W. Goldberg, Phys. Lett. A **127**, 147 (1988)
48. I.A. Min, I. Mezić, A. Leonard, Phys. Fluids **8**, 1169 (1996)
49. K.V. Chukbar, Plasma Phys. Rep. **25**, 77 (1999)
50. E. Ben-Naim, B. Machta, J. Machta, Phys. Rev. E **72** 021302
51. E. Trizac, A. Barrat, M.H. Ernst, Phys. Rev. E **76**, 031305 (2007)
52. A. Hasegawa, K. Mima, M. Duang-van, Phys. Rev. Lett. **54**, 2608 (1985)
53. Z. Meng, R.M. Thorne, D. Summers, J. Plasma Phys. **47**, 445 (1992)
54. C. Tsallis, J. Stat. Phys. **52**, 479 (1988)
55. H. Takagi, M.J. Sato, T. Yanagida, M. Ueda, PLoS One **3**, e2648 (2008)
56. G. Zumofen, J. Klafter, M.F. Shlesinger, in *Anomalous Diffusion. From Basics to Applications*. Lecture Notes in Physics, vol. 519, ed. by R. Kutner, A. Pękalski, K. Sznajd-Weron (Springer, Berlin, 1999), p. 15
57. I. Eliazar, J. Klafter, Ann. Phys. **326**, 2517 (2011)
58. T. Sandev, A. Chechkin, R. Metzler, Anomalous-to-normal diffusive crossover in complex systems: tempered dynamics approach, in *Proceedings of the 12th Conference of the Society of Physicists of Macedonia, 27 – 30 September 2018, Ohrid – Macedonia*, ed. by I. Petreska, O. Galbova, A. Skeparovski, O. Zajkov (Skopje: Society of Physicists of Macedonia, 2019), pp. 10–15

59. V. Sposini, A.V. Chechkin, F. Seno, G. Pagnini, R. Metzler, New J. Phys. **20**(4), 043044 (2018)
60. R. Kutner, M. Regulski, Phys. A **264**, 107 (1999)
61. B. Wang, J. Kuo, S.C. Bae, S. Granick, Nat. Mater. **11**, 481–585 (2012)
62. J.-P. Bouchaud, J. Phys. I (France) **2**, 1705 (1992)
63. G. Bel, E. Barkai, Phys. Rev. Lett. **94**, 240602 (2005)
64. G. Margolin, E. Barkai, Phys. Rev. Lett. **94**, 080601 (2005)
65. Y. He, S. Burov, R. Metzler, E. Barkai, Phys. Rev. Lett. **101**, 058101 (2008)
66. A. Rebenshtok, E. Barkai, Phys. Rev. Lett. **99**, 210601 (2007)
67. A. Lubelski, I.M. Sokolov, J. Klafter, Phys. Rev. Lett. **100**, 250602 (2008)
68. J.-H. Jeon, V. Tejedor, S. Burov, E. Barkai, C. Selhuber-Unkel, K. Berg-Sørensen, L. Oddershede, R. Metzler, Phys. Rev. Lett. **106**, 048103 (2011)
69. A.V. Weigel, B. Simon, M.M. Tankun, D. Krapf, Proc. Natl. Sci. U.S.A. **108**, 6438 (2011)
70. F. Thiel, I.M. Sokolov, Phys. Rev. E **89**, 012136 (2014)
71. R. Kutner, Phys. A **264**, 84 (1999)
72. J. Perelló, J. Masoliver, A. Kasprzak, R. Kutner, Phys. Rev. E **78**, 036108 (2008)
73. A. Kasprzak, R. Kutner, J. Perelló, J. Masoliver, Eur. Phys. J. B **76**, 513 (2010)
74. R. Kutner, Phys. A **314**, 786 (2002)
75. W. Feller, *An Introduction to Probability Theory and its Applications*, vol. 2, 2nd edn. (Wiley, New York, 1971)

# Chapter 5
# A Brief Introduction to DFA-Based Multiscale Analysis

**Paweł Oświęcimka**

**Abstract** In this course, I present the basic theoretical concepts serving a description of the multifractal patterns with focus on their application in the financial time series analysis. I present both advanced algorithms of quantifying the scaling properties of non-stationary data and the fractal models which are extremely useful in modelling complex time series. I also introduce a concept of the multifractal cross-correlation and point to its usefulness in filtering out such effects among the different time series.

## Introduction

Since Benoit Mandelbrot introduced the concept of fractality [60], the interest in such objects has been steadily increasing. This is motivated by a successful identification of fractal structures in many not immediately related domains of scientific research and artistic creation: physics [1, 2], biology [3–5], chemistry [6, 7], economy [8–13], linguistics [14, 15] and even music [16, 17]. A wide scope of this theory and its ability to describe a significant number of varying natural phenomena resulted in it becoming one of the pillars of interdisciplinary research, where barrier crossing is taken as imperative and elevated the fractal theory to the position of a principal tool there.

Fractals are often perceived as complicated structures even if they arise from iteration of a single, simple formula. The key feature of fractals is self-similarity that refers to objects which reiterate themselves at the many levels of resolution (scales). They also differ essentially from the Euclidean geometry's structures when their dimension is considered. Hence, fractals required introduction of the generaliza-

P. Oświęcimka (✉)
Complex Systems Theory Department, Institute of Nuclear Physics Polish Academy of Sciences, PL-31342 Krakow, Poland
e-mail: pawel.oswiecimka@ifj.edu.pl



D. Grech and J. Miśkiewicz (eds.), *Simplicity of Complexity in Economic and Social Systems*, Springer Proceedings in Complexity,
https://doi.org/10.1007/978-3-030-56160-4_5

tion of topological dimension which led to the Hausdorff-Besicovitch concept [18] of fractal dimension. Finally, it appeared that this new geometry describes natural objects very well.

Later this concept led to the recognition of even more complex structures, which are a hierarchical convolution of many fractals (multifractals) [19]. Multifractality has often been perceived in the context of the complex systems ruled by nonlinear interactions. Popularity of this theory benefited a lot from the development of methods identifying such structures in the time series [20, 21]. Developed in the last 20 years, the MFDFA and WTMM algorithms facilitate a quantitative representation of the multifractal properties of the time series with the Hölder exponents distribution function, known also as a multifractal spectrum or the singularity spectrum [22]. Moreover, these methods reveal information not available with the use of the autocorrelation function or the power spectrum function.

Multifractal formalism is also successfully applied to the analysis of financial data. From the point of view of a physicist, financial markets belong to one of the most complex structures in nature [23]. A great number of elements that can be considered complex in themselves and the lattice of interrelations between them imply the use of advanced and non-standard methodology in the analysis of such a system. The first attempt at a mathematical approach to the financial markets was made by Louis Bachelier, who considered the series of price variations in his doctoral thesis of 1900, *Théorie de la Spéculation*, prepared under the supervision of Henri Poincaré. He developed the theory of Brownian motions, which was formally introduced in physics 5 years later by Albert Einstein, and, independently, by Marian Smoluchowski (in 1906). A rigorous theory of Brownian motions was derived by Norbert Wiener in 1923; hence, it is often referred to as the Wiener process. Later, it was revealed that the financial series are more complex than classic Brownian motion, which is an example of monofractal function, and they can be more completely described within the framework of multifractal formalism. Moreover, multifractal analysis of financial time series provides information about the complexity of the system from which they originate and provide valuable information about the correlation structure of these series.

In this article, I present basic concepts of multifractal analysis as well as I introduce the methodology needed for multifractal analysis of time series. Naturally, it is impossible to cover all the aspects of the multifractality in such a short article and that was not my objective. I had rather tried to make a brief introduction to the multifractal analysis. The article structure is as follows. Section 5.2 defines fractal objects and introduces the concept of a fractal dimension. Section 5.3 defines fractal functions (fractional Brownian motion and Weierstrass function) and studies their essential properties. Section 5.4 is devoted to the DFA method. Section 5.5 introduces the concept of multifractality and the model of binomial cascade. Section 5.6 is devoted to the MFDFA methodology. Section 5.7 presents a method of multifractal cross-correlation analysis MFCCA.

## Fractal Structures

Fractals are geometrical objects which escape a precise mathematical definition. Due to the variety of fractal structures, it is difficult to propose a unique definition which can be applied to every fractal. Therefore, fractals are identified through enumeration of characteristic features common for these objects. One of the most basic attributes of the fractals is self-similarity. In common language, it is understood as a similarity of the object fragment, properly rescaled, to the whole structure. This feature is well illustrated on the example of Sierpiński triangle which is one of the well-known fractal mathematical constructions proposed by Polish mathematician Wacław Sierpiński in 1915 (Fig. 5.1a). However, it is worth mentioning that similar objects were also used by ancient architects to decorate flooring an example of which is the Romanesque cathedral in Anagni (Italy) [24]. The Sierpiński triangle can be constructed in many ways [25–27]. In this article, we describe one of the most popular construction based on the reduction rule. We begin our procedure with an equilateral triangle which is divided into four congruent equilateral triangles by connecting midpoints of all sides and the triangle in the centre is removed. This procedure is repeated for each of the smaller triangles and can be continued recursively to infinity. The initial four steps of the algorithm are depicted in Fig. 5.1b. Self-similar properties of the Sierpiński triangle can be qualitatively characterized by an analysis of the elements' distribution in each iteration. Let us denote by $N(L)$ the number of triangles with a side length $L$ for a particular level of the procedure iteration (for initial stage $L = 1$). Thus, $L$ corresponds to characteristics scale or resolution level at which the Sierpiński triangle is analysed. The relation between $N(L)$ and $L$ is given as follows:

$$N(L) = L^{-d_f}, \tag{1}$$

where $d_f$ denotes a fractal dimension. The term fractal dimension was coined by Mandelbrot in [28]; however, the catalyst for this concept was works, among others, of Felix Hausdorff who introduced a mathematically precise definition of the fractional dimension [29, 30]. In a nutshell, dimension is a measure of space-filling properties of the object [31] and it corresponds to the number of degrees of freedom (coordinates) which is needed to be determined to indicate the position of the point in a set. Thus, the topological dimension $d_T$ can take only a positive integer number. However, this standard definition fails when it is applied to a fractal-type object which can be considered as a complex structure with a non-integer dimension. Definition of the dimension proposed by Hausdorff handles this problem and is capable to determine a (non-integer) fractal dimension. Hence, according to the Mandelbrot suggestion, the fractal dimension quantifies the deviation of the structure from its topological ideal [32]. Through rearranging (1), we can obtain a direct formula for $d_f$:

$$d_f = -\frac{log(N(L))}{log(L)}, \tag{2}$$

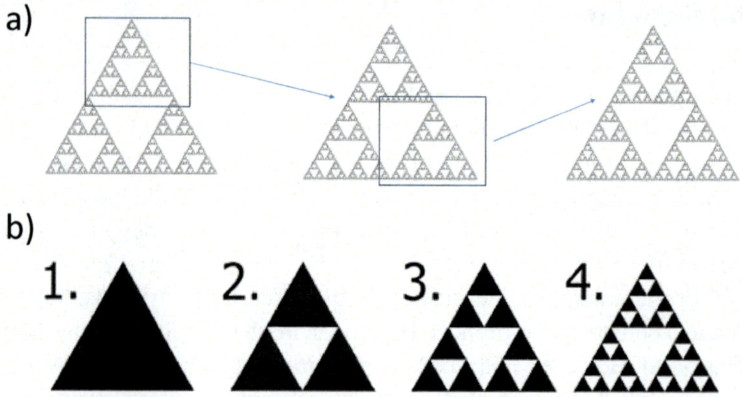


**Fig. 5.1** **a** Consecutive enlargements of Sierpiński triangle which illustrate its self-similar properties. **b** Initial steps of the Sierpiński triangle construction

which in the case of the Sierpiński triangle assumes irrational value ≈1.585. Thus, the fractal dimension is an index which quantitatively describes scaling properties of the analysed structure [20]. $d_f$ can be considered as a generalization of the topological dimension $d_T$ and the following rule is valid:

$$d_T \leq d_f. \tag{3}$$

## Fractal Functions

The concept of fractality is very useful in characterization of a geometrical, irregular object. However, the same formalism can be applied to a quantitative analysis of irregular sharp functions or time series which are investigated in varied scientific research [33]. In this approach, a function can be recognized as fractal if it reveals self-affine properties which means that the function is similar to itself under an anisotropic transformation. A special case of self-affinity is self-similarity related to an isotropic transformation. Thus, the fractal (or self-affine) function $f(x)$ $\forall x$ and $\lambda > 0$ fulfils the following law [20]:

$$f(x_0 + \lambda x) - f(x_0) \simeq \lambda^H f(x_0 + x) - f(x_0), \tag{4}$$

where $H$ denotes the Hurst exponent. The self-similar properties (or isotropic transformation) are retrieved for $H = 1$. An example of an iso- and anisotropic transformation is visually illustrated for the Weierstrass-type function in Fig. 5.2 [34–36]. Weierstrass curve is an interesting example of a log-periodic fractal which is contin-

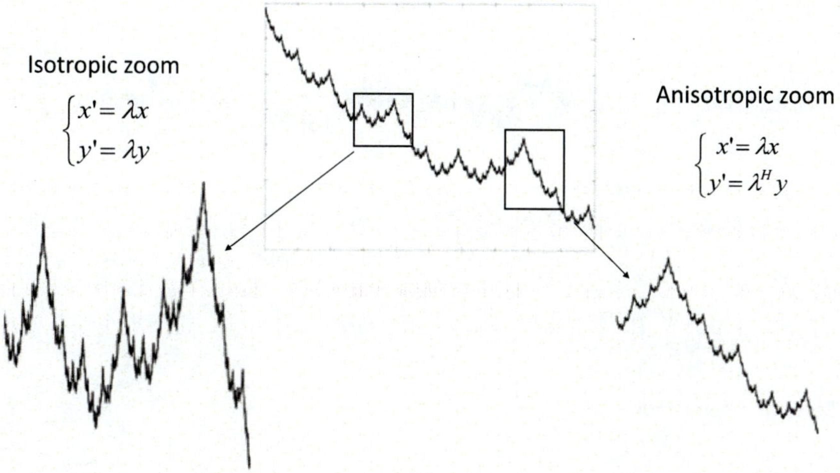


**Fig. 5.2** Different types of transformation of the Weierstrass-type function ($\lambda = 2, \gamma = \pi/2, d_f = 1.4$). Anisotropic zoom refers to self-affine properties of the function

uous everywhere and differentiable nowhere and can be obtained by the expression [37]:

$$f(t) = \sum_{n=0}^{\infty} \frac{1}{\lambda^{(2-d_f)n}} exp(-\lambda^n t cos(\gamma)) cos(\lambda^n t sin(\gamma)), \tag{5}$$

where $\gamma \in [0, \pi/2]$. For $\gamma = \pi/2$ this function develops fractal properties which are characterized by the fractal dimension $d_f$. In this study, we consider the function with $d_f = 1.4$ and $\lambda = 2$.

Another well-known model of the fractal function is the fractional Brownian motion (fBm) originally introduced by Kolmogorov [38] and developed by Mandelbrot and Van Ness [39]. fBm is a generalization of the classical Brownian motion on the case with non-independent increments and has been applied to model phenomena in many different areas in science [40–44]. A Gaussian process with mean value 0 is called fractional Brownian motion $B_t^H$ with the Hurst index $H$ if its covariance function is given in the form [45, 46]:

$$< B_t^H B_s^H >= \frac{\sigma^2}{2} < |t|^{2H} + |s|^{2H} - |t-s|^{2H} > . \tag{6}$$

It is easily observable that for $H = 1/2$ we recover the classical Brownian motion (the covariance function $< B_t^{1/2} B_s^{1/2} >\sim |t| \wedge |s|$ ), and thus increments of this process are independent. For any other value of $H$, the increments are correlated. Among interesting properties of fBm which can be inferred directly from (6) we should

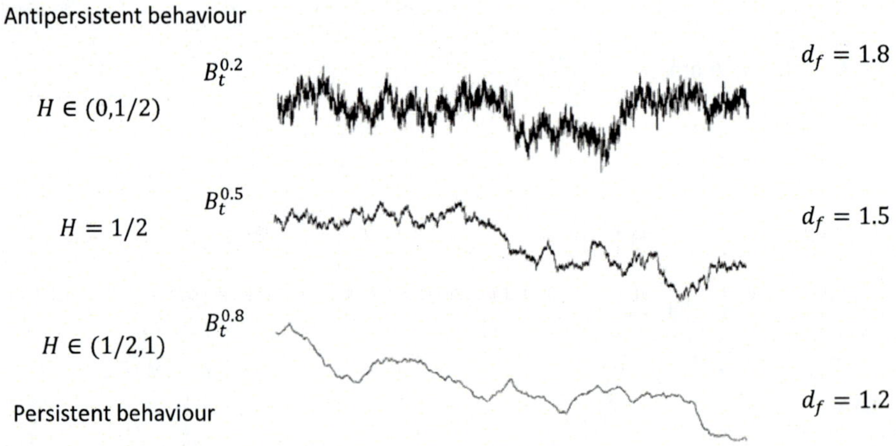


**Fig. 5.3** Fractional Brownian motion generated with different Hurst exponent $H$ and corresponding fractal dimension $d_f$

mention stationarity and self-similarity of its increments. Thus, the distribution of the increments remains the same on different scales $\lambda$ and is provided by the relation:

$$< B^H_{t+\lambda s} - B^H_t > \stackrel{d}{=} \lambda^H < B^H_{t+s} - B^H_t > . \tag{7}$$

This means that the fractional Brownian motion can be considered as a self-affine function with scaling index $H$. Moreover, the regularity of the fBm is entirely determined by the Hurst exponent. The larger the $H$, the more regular the fBm. For $H \in (0, 1/2)$, fBm is antipersistent (covariance is negative), and thus it is more likely to reverse the sign of increments than to keep it. In contrast, for $H \in (1/2, 1)$, fBm is persistent (covariance is positive) and it is more likely to continue the trend (Fig. 5.3). Moreover, the Hurst exponent $H$ and fractal dimension $d_f$ are related according to the equation [47]:

$$d_f = 2 - H. \tag{8}$$

Thus, the larger the Hurst exponent the smaller the fractal dimension and vice versa.

## Analysis of Time Series with the DFA Methodology

One of the reliable method used to estimate the Hurst exponent of time series is the detrended fluctuation analysis (DFA) [48, 49]. This method consists of four steps which are described as follows. Consider time series $x_i$ of length $N$, $i = 1, 2 \ldots N$. Firstly, the so-called profile is calculated according to the formula:

$$X(j) = \sum_{i=1}^{j} [x_i - < x >], \tag{9}$$

where $<>$ denotes average over the entire time series. Since the profile has to be analysed on different time scales, the time series is divided into $N_s$ non-overlapping segments $\nu$ of length $s$ ($N_s = int(N/s)$). Because the length of the time series is not necessarily a multiple of the scale $s$, the procedure of division is also repeated starting from the end of the time series. Hence, we obtain $2N_s$ segments. In each segment, the possible local trend $\nu$ is estimated by fitting polynomial of order $m$ ($P_\nu^{(m)}$) and subsequently subtracted from the data. Thus, the capability of trend elimination strongly depends on the used in the calculation polynomial order [50]. In this study, we used $m = 2$. Next, we calculate the detrended variance in each segment:

$$F^2(\nu, s) = \frac{1}{s} \sum_{k=1}^{s} (X((\nu - 1)s + k) - P_\nu^{(m)}(k)). \tag{10}$$

Finally, the fluctuation function is estimated according to the formula:

$$F(s) = \frac{1}{2N_s} \sum_{\nu=1}^{2N_s} (F^2(\nu, s)). \tag{11}$$

If the analysed time series is fractal, then power-law behaviour is observed as given below:

$$F(s) \sim s^H, \tag{12}$$

where $H$ denotes the Hurst exponent introduced in Sect. 5.3. Therefore, if the time series is short-range correlated, the Hurst exponent assumes value $\sim 0.5$. For long-range correlated time series, the deviation of $H$ from 0.5 is expected. Hence, for $0.5 < H < 1$, the analysed time series is persistent (positively autocorrelated), whereas for $0 < H < 0.5$, the data are antipersistent (negatively autocorrelated). As an example of the analysis by means of the DFA method, we show the results for both increments of the fractional Brownian motion (Fig. 5.4c) as well as sample financial time series representing the Volkswagen company (VOW) listed on the German stock market DAX and comprising the interval between November 28, 1997 and December 31, 1999 (Fig. 5.4d) [11]. In the latter case, the time series of VOW share prices $p(t_i)$, where $i = 1, 2 \ldots, N$ denotes the index of the consecutive transactions, are transformed into two processes: logarithmic price increments $g(i) = ln(p(t_{i+1})) - ln(p(t_i))$ (Fig. 5.4a) and time intervals between consecutive transactions $\Delta t_i = t_{i+1} - t_i$ (Fig. 5.4b) which were subjected to a fractal analysis. In all cases, we observe scaling relations (linear relation in log-log scale) which confirm the fractal character of the studied time series. The estimated Hurst exponents are presented in the corresponding figure. The exponents estimated for fBm correspond to their theoretical counterparts which is an argument in favour of the reliability of

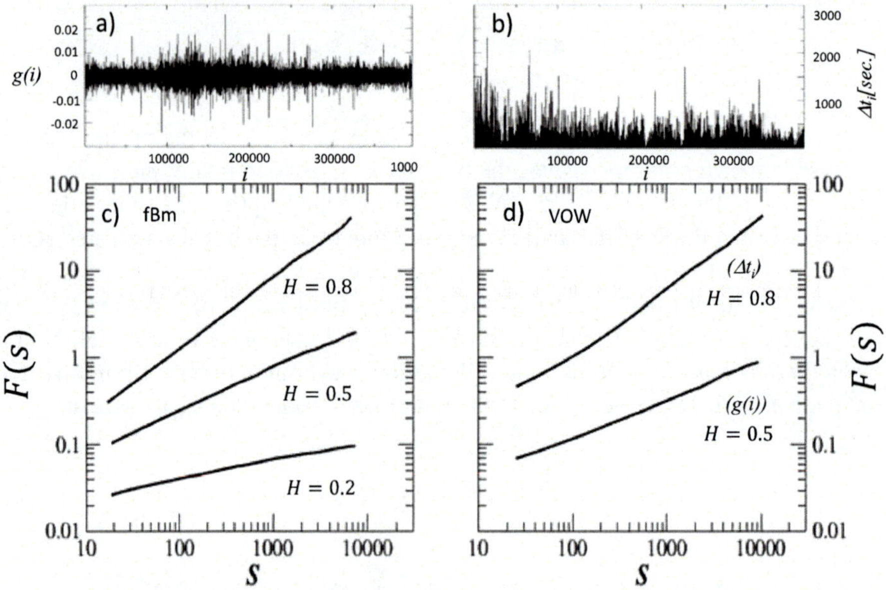


**Fig. 5.4** Plots of the time series of logarithmic price increments (**a**) and time intervals between consecutive transactions (**b**) of the Volkswagen company (VOW). Fluctuation functions $F(s)$ calculated for fBm noise (**c**) as well as price increments and time intervals between consecutive transactions of VOW (**d**)

the DFA algorithm. For the financial time series, we obtain $H$=0.5 for price increments and $H = 0.8$ for waiting times which aptly capture the character of the data: only short-range correlation and persistency for price increments and time interval, correspondingly.

## Multifractality of the Time Series

For complex time series, however, the single scaling exponent is insufficient to characterize the system properly. In this case, the local version of the Hurst exponent called Hölder exponent ($\alpha$) quantifying local regularity of the function is needed. It can be estimated according to the relation [20]:

$$|f(x_0 + \delta) - f(x_0)| \sim C\delta^{\alpha(x_0)}, \tag{13}$$

where $\alpha(x_0)$ determines the singularity strength in $x_0$. Thus, the smaller the exponent $\alpha(x_0)$, the more singular the function in $x_0$ and it is opposite for larger $\alpha(x_0)$. Spectrum of the Hölder exponents is called a singularity (or multifractal) spectrum and is defined as follows:

$$f(\alpha) = d_f(\{x_0, \alpha(x_0) = \alpha\}). \tag{14}$$

Thus, $f(\alpha)$ is a fractal dimension of the data support with particular $\alpha$. The shape of the multifractal spectrum resembles an inverted parabola and its width refers to the time series complexity, i.e. the wider the spectrum, the "richer" the multifractal structure and vice versa.

One of the well-known examples of a multifractal structure is the binomial cascade. There are many variations of this model; however, in this paper, we consider a classical deterministic version [51]. The procedure of generation of the cascade is as follows: we start from the measure which is homogeneously distributed on segment [0,1]. In the first step, the segment is divided into two halves and the initial measure is distributed according to the proportion rate $a$ ($1/2 < a < 1$), i.e. $a$ and $(1 - a)$ portion of the initial measure is assigned to the left and right subintervals, respectively (Fig. 5.5a). Then this process is repeated recursively for each subinterval independently and in theory continues *ad infinitum*. In practice, however, the final iteration level is assumed. At final stage $k$, each subinterval $j$ has size $2^{-k}$ and measure $M_j$ which is a product of $k$ multipliers $a^i_j$ (equals $a$ or $1 - a$) [51]:

$$M_j = \prod_{i=1}^{k} a^i_j, \tag{15}$$

where $j$ denotes a subinterval index and $i$ is the particular stage of the cascade. In this case, the singularity spectrum is derived straightforwardly:

$$\alpha = -\frac{1}{ln(2)} \frac{a^q ln(a) + (1-a)^q ln(1-a)}{a^q + (1-a)^q}, \tag{16}$$

$$f(\alpha) = -\frac{q}{ln(2)} \frac{a^q ln(a) + (1-a)^q ln(1-a)}{a^q + (1-a)^q} - \frac{-ln(a^q + (1-a)^q)}{ln(2)}, \tag{17}$$

where $q \in \Re$. Thus, scaling properties of the cascade are entirely determined by the rate of measure distribution $a$. In Fig. 5.5b, c, we depicted both a binomial cascade with $k = 17$ and $a = 0.65$ and their singularity spectrum, respectively.

An accurate recognition of the multifractal properties in the time series challenges the modern methods of data analysis. This results from the non-stationary character of the experimental data, which may lead to erroneous outcomes when standard methods based on the partition function are applied [52]. Therefore, two competing methods of multifractal analysis which handle the above-mentioned difficulty were proposed, i.e. multifractal detrended fluctuation analysis (MFDFA) [21] and wavelet transform modulus maxima (WTMM) [20]. The first extends the DFA algorithm [48] into the multifractal cases and facilitates assessment of the $f(\alpha)$ singularity spectrum by means of the scaling of the q-th-order moments. The second utilizes a continuous wavelet transform which provides a representation of the fractal structure in the time-frequency plane, and then quantifies the singularity spectrum by means of the

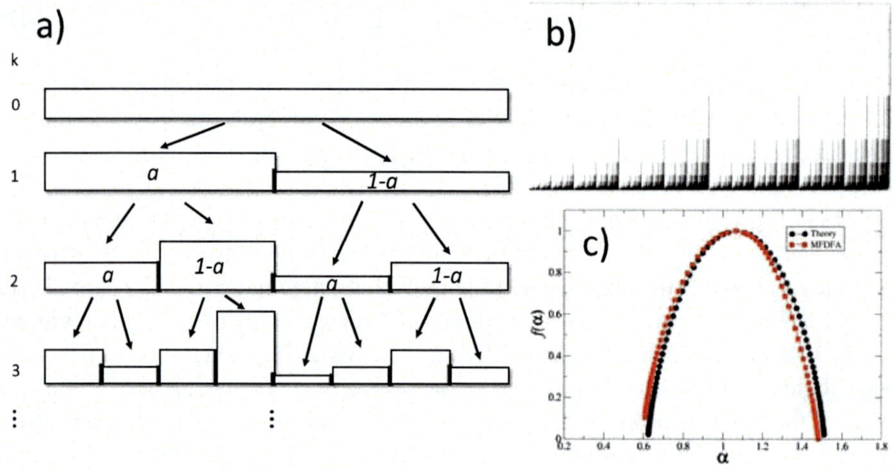


**Fig. 5.5** **a** Visualization of the initial stages of binomial cascade construction. **b** Binomial measure after $k = 17$ stages ($a = 0.65$). **c** Theoretical and estimated through MFDFA method singularity spectrum of the corresponding binomial cascade

properly defined partition function. The WTMM makes the analysed series stationary through the use of the wavelet orthogonal to the low-order polynomials while the MFDFA removes the trend component of the signal by means of polynomial with assumed order. These detrending procedures are necessary for an accurate assessment of the singularity strength in non-stationary signals. In this course, we focus on the MFDFA method.

## Estimation of Multifractality with the MFDFA

Multifractal detrended fluctuations analysis (MFDFA) is a generalization of the DFA algorithm on multifractal case [21]. Therefore, the first steps of the MFDFA leading to estimating the detrended variance $F^2(\nu, s)$ ((10)) are the same as in DFA. Next, the $q$-order fluctuation function must be calculated according to the equation:

$$F_q(s) = \left\{ \frac{1}{2N_s} \sum_{\nu=1}^{2N_s} [(F^2(\nu, s))]^{q/2} \right\}^{1/q}, q \in \Re \setminus \{0\}, \tag{18}$$

where $q$ serves as a filter which discriminates fluctuations with respect to variance amplitude. Thus, for positive $q$ large fluctuations are favoured, whereas for negative $q$ the segments with a small variance are filtered out. In this paper, we used $q \in [-4, 4]$. For fractal, time series $F_q(s)$ reveals power-law behaviour:

$$F_q(s) \sim s^{h(q)}, \tag{19}$$

where $h(q)$ is the generalized Hurst exponent. Thus, scaling properties of fluctuations with different amplitudes correspond to $h$ with different $q$. If the analysed signal is monofractal, $h(q)$ is independent of $q$ and equals the Hurst exponent $h(q) = H$. For multifractal signals, $h(q)$ is a decreasing function of $q$. The spectrum of the generalized Hurst exponent can be converted into a multifractal spectrum through the following formulas:

$$\alpha = h(q) + qh^{'}(q), \quad f(\alpha) = q[\alpha - h(q)] + 1, \tag{20}$$

where $\alpha$ is Hölder exponent and $f(\alpha)$ refers to the fractal dimension of the data support with particular $\alpha$ (see Sect. 5.5). The degree of the complexity of the signal is quantitatively described by the width of the multifractal spectrum $\Delta\alpha = \alpha_{max} - \alpha_{min}$. The wider the spectrum, the greater the complexity of the time series, and the more developed the examined multifractal.[1] As it was shown in Sect. 5.5, for the classic binomial model, the $f(\alpha)$ spectrum is a symmetric inverted parabola. However, the multifractal spectrum of the real-time series usually is not symmetric, having one side better developed than the other one. Therefore, the spectrum asymmetry can carry key information about the temporal organization of the time series. The $f(\alpha)$ spectrum asymmetry results from the heterogeneous character of the process responsible for the fluctuations of varied strength. The asymmetry parameter is defined as [33] follows:

$$A_\alpha = (\Delta\alpha_L - \Delta\alpha_R)/(\Delta\alpha_L + \Delta\alpha_R), \tag{21}$$

where $\Delta\alpha_L$ and $\Delta\alpha_R$ denote the distance from the maximum of the spectrum to, respectively, the smallest and the largest determined value of $\alpha$ and $|A_\alpha|$ defines the degree of asymmetry (absolute value of $A_\alpha$ grows with the increasing asymmetry) while the sign $A_\alpha$ indicates whether the asymmetry is left-sided (positive) or right-sided (negative). In the case of left-sided asymmetry, more developed multifractal properties feature large fluctuations, while small ones are governed by simpler dynamics. This behaviour is revealed by the variations of stock market indices and may be related to the signal in which noise dominates over the small signal amplitude and creates the background for the larger, nonlinearly correlated fluctuations. As an example, the analysis of daily returns of stock market index DAX recorded in the period between 12 January 1990 and 12 October 2013 (5881 data points) (Fig. 5.6a) is presented. The left-sided asymmetry with $A_\alpha = 0.56$ is clearly visible. Moreover, the heterogeneity of the scaling properties can also be noticed in the distribution of the corresponding $F_q(s)$ (Fig. 5.6a inset) for which spread of the fluctuation functions for $q > 0$ is much larger than for the negative $q$.

A less intuitive and less frequent in nature reverse arrangement occurs. The right-sided asymmetry is built when the simpler dynamics govern the large fluctuations while the small ones are more complex in terms of their organization. This kind of singularity spectrum is featured by the intertransaction times on the stock market,

[1]The bifractal time series [53] of two points $f(\alpha)$ spectrum make the exception.

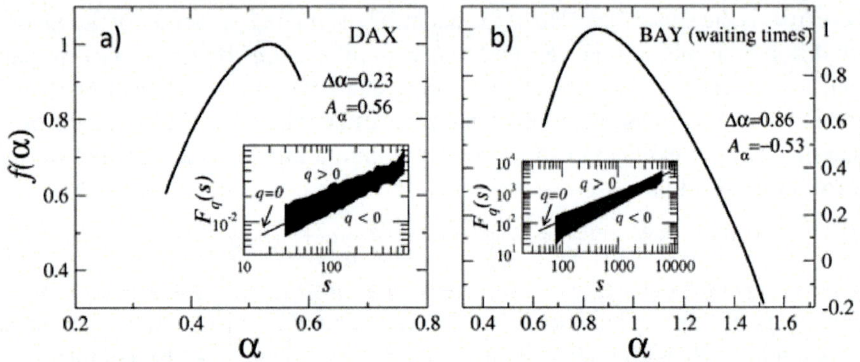


**Fig. 5.6** Singularity spectra $f(\alpha)$ and corresponding fluctuation functions $F_q(s)$ for both DAX index (**a**) and intertransaction time intervals (waiting time) of Bayer company (**b**)

where the small fluctuations indicate an increased activity and greater volatility of the market. An example of the corresponding multifractal spectrum estimated for intertransaction times of the Bayer (BAY) company over the period between 28 November 1997 and 31 December 1999 is depicted in Fig. 5.6b. This asymmetry in organization of the fluctuations is also visible at the $F_q(s)$ level for which the spread of the fluctuation function for $q < 0$ is much larger than for $q > 0$. Intriguingly, this kind of asymmetry was also identified in fluctuations in number of the solar spots computed by the well-known Wolf formula [33].

## Multifractal Cross-Correlation Analysis

Multifractal analysis of the time series proved to be a useful method of description of the nonlinear relations in complex systems' signals. However, in many situations, the analysis of cross-correlation between time series representing different observables is crucial for a proper characterization of the system. Moreover, similarly to the case of the autocorrelation analysis, cross-correlation can also exhibit nonlinear properties and be susceptible to non-stationarity of the time series. These problems, as it was presented above, were overcome within the multifractal formalism. Therefore, expansion of this methodology on the cross-correlation case, which could describe fractal relations between distinct signals, seems natural. This idea was especially intensively developed within the DFA algorithm; although other methods, e.g. based on wavelet approach, were also proposed [54]. The first step in this direction has been taken with the proposition of the detrended cross-correlation analysis (DCCA) [55], which generalizes the well-known DFA method to the two series case. The DCCA method is able to identify fractal cross-correlation between simultaneously recorded time series and quantifies it by the scaling exponent $\lambda$. However, cross-correlation for the complex systems needs to be quantified by a set of scaling exponents retrieved from the covariance function [56] rather than a single one. It was the motivation to introduce the multifractal detrended cross-correlation analysis method (MF-DXA) [57]. However, this algorithm did not offer a reliable estimation of the scaling exponents because of the properties of the covariance function used. Taking the modulus of the covariance within the framework of the procedure leads to the detection of the fractal cross-correlation between independent signals. A proper generalization of the DCCA algorithm on multifractal case is the multifractal cross-correlation analysis (MFCCA) [58] where higher order moments are constructed with the preservation of the sign of the covariance and thus facilitates the estimation of the scaling exponents $\lambda_q$. Algorithm of the MFCCA can be briefly described in the following steps.

Consider two time series $x_i$, $y_i$ where $i = 1, 2 \ldots N$. At the first step, the signal profile must be calculated for each of them:

$$X(j) = \sum_{i=1}^{j} [x_i - <x>], \quad Y(j) = \sum_{i=1}^{j} [y_i - <y>], \tag{22}$$

where $<>$ denotes averaging over the entire time series. Then, these profiles have to be divided into $2M_s$ ($M_s = int(N/s)$) disjointed segments $\nu$ of length $s$ starting from the beginning as well as the end of the time series. For each box $\nu$, the assumed trend ($P^{(m)}_{X,\nu}$ for $X$ series and $P^{(m)}_{Y,\nu}$ for $Y$) is estimated by fitting a polynomial of order $m$. In this paper, we used $m = 2$. Next, the trend is subtracted from the data and the detrended cross-covariance within each box is calculated:

$$F^2_{xy}(\nu, s) = \frac{1}{s} \sum_{k=1}^{s} \{(X((\nu-1)s+k) - P^{(m)}_{X,\nu}(k))(Y((\nu-1)s+k) - P^{(m)}_{Y,\nu}(k))\}. \tag{23}$$

Next, to amplify (suppress) the values of covariance function, the q-filtering technique is applied and the $q$th-order covariance function is calculated by the following equation:

$$F^q_{xy}(s) = \frac{1}{2M_s} \sum_{k=1}^{2M_s} \text{sign}(F^2_{xy}(\nu, s)) \cdot |F^2_{xy}(\nu, s)|^{q/2}, \quad q \in \Re \setminus \{0\}, \tag{24}$$

where $\text{sign}(F^2_{xy}(\nu, s))$ stands for the sign of the $F^2_{xy}(\nu, s)$ function. Thus, information about genuine signs of $F^2_{xy}(\nu, s)$ is also preserved in $F^q_{xy}(s)$. For $q = 0$, the logarithmic version of (24) must be applied [21]:

$$F^0_{xy}(s) = \frac{1}{2M_s} \sum_{k=1}^{2M_s} \text{sign}(F^2_{xy}(\nu, s)) \cdot ln|F^2_{xy}(\nu, s)|. \tag{25}$$

The $F^q_{xy}(s)$ has to be calculated for many different scales $s$. For fractal cross-correlated time series, the power-law relation between $F^q_{xy}(s)$ and $s$ is observed and the following relation is fulfilled:

$$F^q_{xy}(s)^{1/q} = F_{xy}(q, s) \sim s^{\lambda_q}, \tag{26}$$

(or $exp(F^0_{xy}(s)) = F_{xy}(0, s) \sim s^{\lambda_0}$ for $q = 0$) where $\lambda_q$ is an exponent that quantitatively characterizes the fractal properties of the cross-covariance. For fractally cross-correlated time series, $\lambda_q$ is independent of $q$ and equals $\lambda$ retrieved from the DCCA algorithm. In the case of multifractal cross-correlation, however, $\lambda_q$ is monotonically decreasing function of $q$ and $\lambda_2 = \lambda$.

Practical application of this method is demonstrated on the artificially generated cross-correlated series as well as on the stock market signals. To generate multifractally cross-correlated time series, the Markov-switching multifractal model (MSM) can be applied [59]. It is an iterative algorithm able to generate a multifractal cascade and self-similar structures. It is also often used to simulate natural multifractal time series, an example of which is financial data. In MSM, the returns are modelled as

$$r_t = \sigma_t u_t, \tag{27}$$

where $u_t$ denotes innovations drawn from a standard normal distribution $N(0,1)$ and $\sigma_t$ stands for instantaneous volatility component represented by a multifractal structure. Thus, the volatility component is a product of $k$ multipliers $M_1(t)$, $M_2(t)$, ..., $M_k(t)$ and constant factor $\sigma^2$ according to the equation:

$$\sigma_t^2 = \prod_{i=1}^{k} M_i(t). \tag{28}$$

Each multiplier can be renewed at time $t$. However, to keep the hierarchical structure of the construction, the transition probability $\gamma_i$ depends on the multiplier rank:

$$\gamma_i = 1 - (1 - \gamma_k)^{b^{i-k}}, \tag{29}$$

with $\gamma_k \in [0, 1]$ and $b \in (0, \infty)$. Thus, the multiplier is renewed with probability $\gamma_i$ and remains unchanged with probability $1 - \gamma_i$. The model also needs specification of the multiplier distribution. In this paper, the binomial distribution is used, and thus the multiplier can take value $m_0$ and $2 - m_0$ $(1 \leq m_0 \leq 2)$ with equal probability. Moreover, $\gamma_k = 0.5$ and $b = 2$ parameters are chosen which leads to the following relation of the transition probability:

$$\gamma_i = 1 - (0.5)^{2^{i-k}}. \tag{30}$$

The cross-correlation between different realizations of the model is enforced by the preservation of the same hierarchical structure in each of the series. Differences between the generated series result from the varying values of $m_0$. In this study, we present results for two pairs of signals with $m_0 = \{1.2, 1.35\}$ for the first and

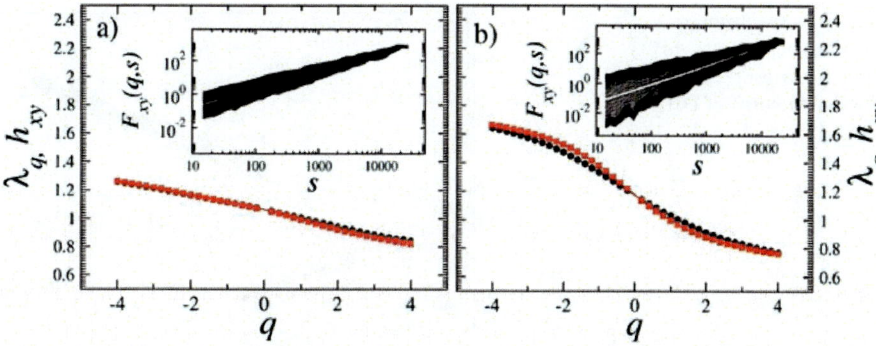


**Fig. 5.7** Spectrum of the cross-correlation exponents $\lambda_q$ and corresponding scaling functions $F_{xy}(q, s)$ for the two pairs of the signals with $m_0 = \{1.2, 1.35\}$ (**a**) and $m_0 = \{1.2, 1.6\}$ (**b**)

$m_0 = \{1.2, 1.6\}$ for the second pair. For both pairs, the MFCCA analysis is performed and the corresponding scaling exponents are estimated. Figure 5.7 shows the scaling functions $F_{xy}(q, s)$ for both cases. They reveal almost perfect multiscaling behaviour which is particularly clear for the pair with the larger difference between $m_0$ parameters (Fig. 5.7b). The resulting spectrum of $\lambda_q$ as a function of $q$ confirms the multifractal character of the correlation between the series. Monotonically decreasing $\lambda_q$ is characteristic for the multifractal structure and the rate of its variations corresponds to the degree of multifractal cross-correlation. In the presented case, a more complex cross-correlation is attributed to the signals pair with $m_0 = 1.6$. To assess the strength of the coupling between the analysed time series, $\lambda_q$ spectrum must be compared with an average of the generalized Hurst exponents calculated as follows [58]:

$$h_{xy}(q) = [h_x(q) + h_y(q)]/2, \tag{31}$$

where $h_x(q)$ and $h_y(q)$ correspond to the fractal properties of individual time series. For the strongest cross-correlation, the $\lambda_q$ and $h_{xy}$ are equal. In the analysed cases, the difference between $\lambda_q$ and $h_{xy}(q)$ is negligible which confirms strong correlations between the time series.

As an example of an analysis of financial time series, the cross-correlation between daily returns of two world leading stock markets DJIA and DAX is investigated. In this study, we consider the indices in the period between 12 January 1990 and 12 October 2013. The corresponding $F_{xy}(q, s)$ characteristics and spectrum of $\lambda_q$ are shown in Fig. 5.8. In contrast to the MSM results, the multifractal cross-correlation is identified only for the positive $q$ in the presented case. This means that the large fluctuations mostly prove to be the carrier of these correlation while small events are much more independent. Moreover, the departure of $\lambda_q$ from $h_{xy}(q)$ indicates that the coupling between indices is not as strong as in the case of the above-considered MSM time series.

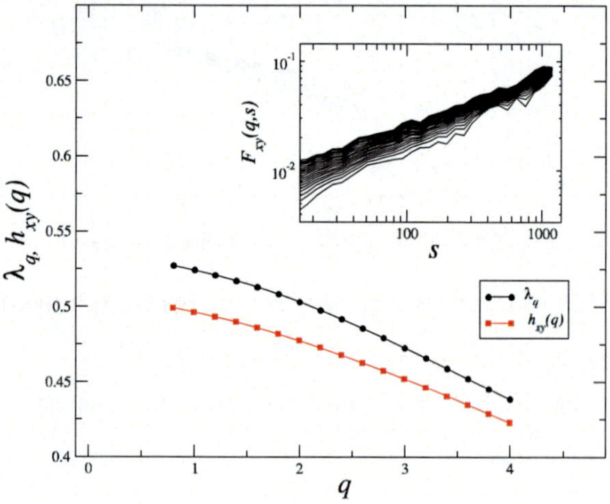


**Fig. 5.8** Spectrum of $\lambda_q$ (main panel) and $F_{xy}(q, s)$ functions (inset) estimated for daily returns of DJIA and DAX indices

## Summary

Many natural phenomena and systems can quantitatively be described within the framework of the fractal geometry which brings such characteristics like the fractal dimension, scaling exponents and cascade effects. This interdisciplinary concept is especially useful for grasping the principal characteristics of time series representing measurements of observables coming from complex systems. For such complex signals, the standard analytic tools, which are sensitive mostly to linear dependencies in a signal, cannot completely describe the real dynamics of the system. In contrast, the concept of multiscale analysis proved viable here. For instance, the multifractal approach facilitates the quantitative description of both linear and nonlinear correlations in the signal. In this paper, the DFA-based techniques of multiscale analysis of complex time series are presented. The capabilities of the multifractal analysis are demonstrated on both sample financial time series and artificially generated data. That study demonstrates that multifractal formalism formed a new, effective way of analysis of the empirical data pertaining to the process of high complexity and varying statistical properties. Thus, it may be applied to various domains of science.

## References

1. J.F. Muzy, E. Bacry, R. Baile, P. Poggi, Uncovering latent singularities from multifractal scaling laws in mixed asymptotic regime. Application to turbulence. EPL(Europhys. Lett.) **82**, 60007 (2008)
2. A.R. Subramaniam, I.A. Gruzberg, A.W.W. Ludwig, Boundary criticality and multifractality at the two-dimensional spin quantum Hall transition. Phys. Rev. B **78**, 245105 (2008)
3. PCh. Ivanov, L.A.N. Amaral, A.L. Goldberger, S. Havlin, M.G. Rosenblum, Z.R. Struzik, H.E. Stanley, Multifractality in human heartbeat dynamics. Nature **399**, 461–465 (1999)

4. A. Rosas, E. Nogueira Jr., J.F. Fontanari, Multifractal analysis of DNA walks and trails. Phys. Rev. E **66**, 061906 (2002)
5. D. Makowiec, A. Dudkowska, R. Gałąska, A. Rynkiewicz, Multifractal estimates of monofractality in RR-heart series in power spectrum ranges. Phys. A **388**, 3486–3502 (2009)
6. H.E. Stanley, P. Meakin, Multifractal phenomena in physics and chemistry. Nature **335**, 405–409 (1988)
7. V.V. Udovichenko, P.E. Strizhak, Multifractal properties of copper sulfide film formed in self-organizing chemical system. Theor. Exp. Chem. **38**, 259–262 (2002)
8. M. Ausloos, K. Ivanova, Multifractal nature of stock exchange prices. Comput. Phys. Commun. **147**, 582–585 (2002)
9. A. Turiel, C.J. Perez-Vicente, Role of multifractal sources in the analysis of stock market time series. Phys. A **355**, 475–496 (2005)
10. R. Kutner, M. Ausloos, D. Grech, T. Di Matteo, C. Schinckus, H.E. Stanley, Econophysics and sociophysics: their milestones and challenges. Phys. A **516**, 240–253 (2019)
11. P. Oświęcimka, J. Kwapień, S. Drożdż, Multifractality in the stock market: price increments versus waiting times. Phys. A **347**, 626–638 (2005)
12. Y.-P. Ruan, W.-X. Zhou, Long-term correlations and multifractal nature in the intertrade durations of a liquid Chinese stock and its warrant. Phys. A **390**, 1646–1654 (2011)
13. D. Grech, G. Pamuła, Multifractality of nonlinear transformations with application in finances. Acta Phys. Pol. A **123**, 529–537 (2013)
14. M. Ausloos, Generalized Hurst exponent and multifractal function of original and translated texts mapped into frequency and length time series. Phys. Rev. E **86**, 031108 (2012)
15. S. Drożdż, P. Oświęcimka, A. Kulig, J. Kwapień, K. Bazarnik, I. Grabska-Gradzińka, J. Rybicki, M. Stanuszek, Quantifying origin and character of long-range correlations in narrative texts. Inf. Sci. **331**, 32–44 (2016)
16. G.R. Jafari, P. Pedram, L. Hedayatifar, Long-range correlation and multifractality in Bach's Inventions pitches. J. Stat. Mech.: Theory Exp. **2007**, P04012 (2007)
17. P. Oświęcimka, J. Kwapień, I. Celińska, S. Drożdż, R. Rak, Computational approach to multifractal music. arXiv:1106.2902
18. H.-O. Peitgen, D.H. Richter, *The Beauty of Fractals: Images of Complex Dynamical Systems* (Springer, New York, 1986)
19. A. Bunde, S. Havlin, *Fractals and Disordered Systems* (Springer, New York, 1991)
20. J.F. Muzy, E. Bacry, A. Arneodo, The multifractal formalism revisited with wavelets. Int. J. Bifurc. Chaos **4**, 245 (1994)
21. J.W. Kantelhardt, S.A. Zschiegner, E. Koscielny-Bunde, S. Havlin, A. Bunde, H.E. Stanley, Multifractal detrended fluctuation analysis of nonstationary time series. Phys. A **316**, 87(2002)
22. T.C. Halsey, M.H. Jensen, L.P. Kadanoff, I. Procaccia, B.I. Shraiman, Fractal measures and their singularities: the characterization of strange sets. Phys. Rev. A **33**, 1141 (1986)
23. J. Kwapień, S. Drożdż, Physical approach to complex systems. Phys. Rep. **515**, 115–226 (2012)
24. E. Conversano, L. Tedeschini-Lalli, Sierpinski triangles in stone on medieval floors in Rome. APLIMAT J. Appl. Math. **4**, 114–122 (2011)
25. W. Sierpinski, Sur une courbe dont tout point est un point de ramification. Comptes rendus de l'Académie des Sciences **160**, 302–305 (1915)
26. D.P. Feldman, *Chaos and Fractals: An Elementary Introduction* (Oxford University Press, Oxford, 2012)
27. T. Rumpf, Conway's game of life accelerated with openCL, in *Proceedings of the Eleventh International Conference on Membrane Computing (CMC 11)*, pp. 459–462 (2010)
28. B.B. Mandelbrot, *Les Objects Fractals: Forme, Hasard et Dimension* (Flammarion, 1975)
29. F. Hausdorff, Dimension und aüsseres mass. Mathematische Annalen **79**, 157–179 (1918)
30. A.S. Besicovitch, On linear sets of fractional dimensions. Mathematische Annalen **101**, 161–193 (1929)
31. M.J. Ostwald, J. Vaughan, *The Fractal Dimension of Architecture* (Birkhäuser, Basel, 2016)
32. Y.-H. Lee, J.R. Carr, D.J. Barr, C.J. Haas, The fractal dimension as a measure of the roughness of rock discontinuity profiles. Int. J. Rock Mech. Min. Sci. Geomech. Abstr. **27**, 453–464 (1990)

33. S. Drożdż, P. Oświęcimka, Detecting and interpreting distortions in hierarchical organization of complex time series. Phys. Rev. E **91**, 030902(R) (2015)
34. D. Nicolletti, Properties of the weierstrass function in the time and frequency domains. Chaos Solitons Fractals **5**, 1–8 (1995)
35. S. Gluzman, D. Sornette, Log-periodic route to fractal functions. Phys. Rev. E **65**, 036142 (2002)
36. P. Oświęcimka, S. Drożdż, J. Kwapień, A.Z. Górski, Fractals, log-periodicity and financial crashes. Acta Phys. Pol. A **117**, 637–639 (2010)
37. M. Bartolozzi, S. Drożdż, D.B. Leinweber, J. Speth, A.W. Thomas, Self-similar log-periodic structures in western stock markets from 2000. Int. J. Mod. Phys. C **16**, 1347–1361 (2005)
38. A. Kolmogorov, Wienersche Spiralen und einige andere interessante kurven in Hilbertschen raum. C.R. (Doklady) Acad. Sci. USSR (N.S.) **26**, 115–118 (1940)
39. B.B. Mandelbrot, J.W. Van Ness, Fractional Brownian motions, fractional noises and applications. SIAM Rev. **10**, 422–437 (1968)
40. P. Cheridito, Arbitrage in fractional Brownian motion models. Financ. Stoch. **7**, 533–553 (2003)
41. L.C.G. Rogers, Arbitrage with fractional Brownian motion. Math. Financ. **7**, 95–105 (1997)
42. P. Chigansky, M. Kleptsyna, Exact asymptotics in eigenproblems for fractional Brownian covariance operators. Stoch. Process. Their Appl. **128**, 2007–2059 (2018)
43. I. Norros, On the use of fractional Brownian motion in the theory of connectionless networks. IEEE J. Sel. Areas Commun. **13**, 953–962 (1995)
44. W.E. Leland, M.S. Taqqu, W. Willinger, D.V. Wilson, On the self-similar nature of ethernet traffic. ACM SIGCOMM Comput. Commun. Rev. **23**, 183–193 (1993)
45. G. Shevchenko, Fractional Brownian motion in a nutshell. Int. J. Mod. Phys.: Conf. Ser. **36**, 1560002 (2015)
46. P. Flandrin, On the spectrum of fractional Brownian motions. IEEE Trans. Inf. Theory **35**, 197 (1989)
47. B.B. Mandelbrot, Self-affinity and fractal dimension. Phys. Scr. **32**, 257–260 (1985)
48. C.-K. Peng, S.V. Buldyrev, S. Havlin, M. Simons, H.E. Stanley, A.L. Goldberger, Mosaic organization of DNA nucleotides. Phys. Rev. E **49**, 1685 (1994)
49. J.W. Kantelhardt, E. Koscielny-Bunde, H.H.A. Rego, S. Havlin, A. Bunde, Detecting long-range correlations with detrended fluctuation analysis. Phys. A: Stat. Mech. Its Appl. **295**, 441 (2001)
50. P. Oświęcimka, S. Drożdż, J. Kwapień, A.Z. Górski, Effect of detrending on multifractal characteristics. Acta Phys. Pol. A **123**, 597–603 (2013)
51. T. Lux, The markov-switching multifractal model of asset returns: GMM estimation and linear forecasting of volatility. J. Bus. Econ. Stat. **26**, 194–210 (2008)
52. J.F. Muzy, E. Bacry, A. Arneodo, Multifractal formalism for fractal signals: the structure-function approach versus the wavelet-transform modulus-maxima method. Phys. Rev. E **47**, 875 (1993)
53. H. Nakao, Multi-scaling properties of truncated Lévy flights. Phys. Lett. A **266**, 282–289 (2000)
54. Z.-Q. Jiang, X.-L. Gao, W.-X. Zhou, H.E. Stanley, Multifractal cross wavelet analysis. Fractals **25**(6), 1750054 (2017)
55. B. Podobnik, H.E. Stanley, Detrended cross-correlation analysis: a new method for analyzing two nonstationary time series. Phys. Rev. Lett. **100**, 084102 (2008)
56. R. Rak, S. Drożdż, J. Kwapień, P. Oświęcimka, Detrended cross-correlations between returns, volatility, trading activity, and volume traded for the stock market companies. EPL(Europhys. Lett.) **112**, 48001 (2015)
57. W.-X. Zhou, Multifractal detrended cross-correlation analysis for two nonstationary signals. Phys. Rev. E **77**, 066211 (2008)
58. P. Oświęcimka, S. Drożdż, M. Forczek, S. Jadach, J. Kwapień, Detrended cross-correlation analysis consistently extended to multifractality. Phys. Rev. E **89**, 023305 (2014)
59. R. Liu, T. Di Matteo, T. Lux, True and apparent scaling: the proximity of the Markov-switching multifractal model to long-range dependence. Phys. A **383**, 35–42 (2007)
60. B.B. Mandelbrot, *The Fractal Geometry of Nature* (Freeman, San Francisco, 1982)

# Chapter 6
# Complex Dynamics of Economic Models with Time Delay

**Marek Szydłowski and Adam Krawiec**

**Abstract** The main aim of this paper is put some economic models in context of complex systems theory. The analogy between economic and physical systems is worth to explore especially in context of oscillating phenomena and dynamical complexity.

## Introduction

### *Complexity in Nonlinear Dynamics of Economic Models*

The main aim of this paper is put some economic models in context of complex systems theory. The analogy between economic and physical systems is worth to explore especially in context of oscillating phenomena and dynamical complexity. We present some economic models formulated in terms of dynamical systems with time delay. Such systems appear in physics and other sciences contexts [1]. We show that oscillatory behavior found in economic systems can be described in an analogous way to he Lienard system in terms of self-sustained oscillations. While economics and physics have different domains, we can find in both disciplines analogous models described by differential equations. These equations tell us how dynamical mechanisms work. In consequence one can apply analogous reasoning in economics based

M. Szydłowski (✉)
Astronomical Observatory, Jagiellonian University, Kraków, Poland
e-mail: marek.szydlowski@uj.edu.pl

Mark Kac Complex Systems Research Centre, Jagiellonian University, Kraków, Poland

A. Krawiec
Institute of Economics, Finance and Management, Jagiellonian University, Kraków, Poland

 
D. Grech and J. Miśkiewicz (eds.), *Simplicity of Complexity in Economic and Social Systems*, Springer Proceedings in Complexity,
https://doi.org/10.1007/978-3-030-56160-4_6

on well-known recognized physical mechanisms. It is reflection of interdisciplinary character of dynamical systems methods and complexity investigations.

In an economy we observe very often non-regular types of behaviors. If we consider financial market dynamics, economic growth, price fluctuation or business cycle, then in the phase space emerged complexity in the behavior of trajectories describing changes of micro and macro economic state variables. In the recent years, nonlinear dynamical systems plays crucial role in description of complexity of dynamics.

Arthur has written [2, p. 18]

> Complex systems arise naturally in the economy. Economic agents, whether they are banks, consumers, firms, or investors, continually adjust their market moves, buying decisions, prices, or forecasts together create. But unlike oins in spin glass that reacts damply to their local magnetic field, economic elements (human agents) react with strategy and foresight by considering out comes that may might result as a consequence of behavior they might undertake. This adds a layer of complication to economics not experienced in other sciences (physics or immunology).

Arthur developed a new approach in neoclassical economics inspired by complex systems methods and proposed start from the behavior of a single agent and assume that the aggregate dynamics simply reproduces the this representative agent average dynamics on the larger scale (see [3, p. 17]).

In economics there is a tradition of using dynamical system methods in the context of investigation of dynamics of different economical models in continuous time (Kalecki, Harrod, Kaldor and Goodwin) as well as in discrete time. Piscitelli and Sportelli discovered complex chaotic behavior of the Shilnikov type in the 3D Harrod model [4]. Very important contribution in investigation of dynamics and model formulation in this mathematical language comes from Chiarella and Flaschel [5–7].

The earliest endogenous cycle models were proposed by Nicholas Kaldor, John Hicks, and Richard Goodwin. But by the late 1950's the dominant approach became the Slutsky–Frisch–Tinbergen methodology of exogenous stochastic 'impulses' that are transformed into a characteristic pattern of oscillations through the filtering properties of the economy's 'propagation mechanism'.

This shift in the methodological approach was probably caused by some disadvantages of endogenous cycle models. They were essentially nonlinear while linear specifications of the exogenous shock models were more convenient. The endogenous cycle models gave too regular periodic motion unlike actual business cycle. They have also inadequate behavioral foundations.

Recent work, however, has seen a revival of interest in the hypothesis that aggregate fluctuations might represent an endogenous phenomenon that would persist without stochastic 'shocks' to the economy.

In presence of nonlinearities complex behavior can emerge from quite simple economic structures. Such fluctuations can converge to a cycle of any regular periodicity or exhibit erratic patterns like those observed in reality. Regular behavior is often interspersed with irregular one. Moreover, for the latter, errors of estimation in initial conditions are accumulated exponentially into substantial errors of forecast.

For example, it was demonstrated that optimal capital accumulation may exhibit irregular behavior [8].

The models will be studied by using the methods of the dynamical systems theory. It allows to study in qualitative way the dynamics of system. The methods of this theory make possible to study the existence the steady states of the system but also the various kind of cyclical or irregular behavior, namely limit cycles and deterministic chaos [9]. The bifurcation theory and deterministic chaos theory is a very convenient way to study the complex behavior in simple, low-dimensional dynamical systems.

This research is based on the assumption that the dynamics of an economy is induced by nonlinear mechanisms. Nonlinearities play the crucial role in generating the variety of dynamic behavior. Different types of periodic and quasi-periodic as well as chaotic fluctuations can be obtained by means of purely deterministic mechanisms. Cyclic phenomena in an economy might be caused by nonlinearities. It seems that the growth cycles are a good example to study this hypothesis. The main task is to find a convincing nonlinear mechanism that is capable of generating the irregular growth paths.

The application of nonlinear dynamics in the context of growth theory were done by Nishimura and Yano [10]. They demonstrated the possibility of ergodically chaotic optimal accumulation for specific discount of future utility. The continuous-time models were studied by Medio [11].

One of the most recent is Sportelli's paper [12] and Asea and Zak's paper [13]. We are also going to use the mathematical methods of functional differential equations. Especially I want to introduce the time lag in production (as Asea and Zak) instead of Nishimura and Yano's condition on the utility to obtain irregular, chaotic growth path. In this way we will obtain the more realistic mechanism of endogenous fluctuations.

Economic development also depends on the properties of investment process. Capital accumulation can be very complex due to a time-to-build technology.

Models with time lags incorporate a delay during which capital is produced, delivered and installed. These models are essentially general equilibrium versions of Kalecki's linear lag model [14]. They have the ability to generate endogenous cycles and thus are free from the debate of the source of shocks.

The bifurcation theory is a major part of research in dynamical economics. A central topic to this is the question whether the qualitative properties of a dynamical system change when one or more of the exogenous parameters are changing. The Hopf bifurcation leads from a critical point to a limit cycle. The bifurcation scenario is one of the way to chaotic behavior.

The system of time delay differential equations is equivalent to the infinite dimensional system of ordinary differential equations. Such systems can exhibit cyclic behavior, for example a limit cycle after the Hopf bifurcation, or even chaotic dynamics. The methods of this theory can be used to analyze economic models with different time lags.

The functional analysis and bifurcation theory offer a few techniques to deal with dynamical problems [15]. In the detection of the limit cycle behavior there are three fundamental methods.

Firstly, one can prove the existence of the limit cycle due to the Poincaré-Bendixson theorem. This line was represented by Chang and Smyth [16] in their study of the Kaldor model. Secondly, in the proof of existence of limit cycle one can refer to the Liénard or Levinson-Smith theorems, as was done by Ichimura [17]. At last, one can use the Poincaré–Andronov–Hopf theorem which also works for the functional differential equations.

This paper follows the third strategy. The detailed analysis of the Hopf bifurcation is performed and the characteristics of the cycle are determined in dependence on the model parameters. It is shown that the cyclic behavior is generated by the time-to-build parameter for generic values of parameters.

## *Oscillations in Economic Models*

One of the founders of econometrics, Norwegian economist Ragnar Frisch was convinced that economic systems are stable and some cyclic phenomena are only the result of exogenous shocks. His influence on economists working in mathematical economics domain made that the paper "Les systèmes autoentretenus et les oscillations de relaxation" published in 1933 went mostly unnoticed [3]. In this paper French mathematician, engineer and physicist Philippe Le Corbeiller argued that economists should use non-linear dynamics methods from Van der Pol to explain the cyclic behavior in economic systems. A Van der Pol oscillator which appeared in an electronic analysis was a candidate for a mechanism to model the business cycle in an economy.

While the mathematical economics followed mostly the path indicated by Frisch and others, the non-linear models of economic started to appear. Le Corbeiller's argument can be find implicit seminal economic works by Kalecki, Kaldor, Goodwin. In the early 1940s Richard Goodwin met Le Corbeiller at Harvard and start to formulate the first model of the business cycles inspired by nonlinear methods used in physics. It was published in Econometrica in 1951 where Goodwin showed a business cycle model represented by the Lienard equation in which a stable limit cycle can appear naturally [18].

While the Kalecki model of business cycle is linear and a limit cycle type of behavior cannot appear because the system is linear at very beginning, this model incorporates time delay in delay which can lead to simple cyclic solution. His pioneering work on time lags in economics was inspiration for including delays in non-linear economic models for modeling cyclic behavior of many economic processes.

In the detection of dynamical complexity the important role can play cyclic behaviors because they can be precursor of complex chaotic behavior like in Ruelle scenarios of transitions to chaos.

Our idea is looking for different sources of cyclicity in economic models in the framework of delay differential equation modeling the dynamics of macroeconomic models.

As the examples will be presented the Kaldor–Kalecki business cycle model [19] and the Kaldor–Kalecki growth model [20]. The former was extensively studied for the existence of bifurcations and stability of a limit cycle [21–25].

Although we will discuss the deterministic version of the Kaldor–Kalecki model it is necessary to mention the the stochastic Kaldor–Kalecki model of business cycle. Due to stochastic perturbation some new dynamic scenarios were found [26]. In model with noise some new criteria ensuring stochastic stability were found by analyzing the Lyapunov exponent, as well as P-bifurcation and pitchfork bifurcation were studied [27].

Both Kaldor–Kalecki business cycle and economic growth models are formulated as systems of functional differential equations.

## *Functional Differential Equations*

The dynamical system theory has its roots in Poincaré work at the end of the nineteenth century [28]. The dynamical system is given by the system of ordinary differential equations

$$\frac{\mathrm{d}x^i}{\mathrm{d}t} = F(x_1, x_2, \ldots, x_n), \quad i = 1, \ldots, n, \tag{1}$$

where $F(x^i)$ are $C^\infty$ functions.

The aim of qualitative analysis of dynamical systems is to determine, for all initial conditions, all types of solutions. It means to find critical points ($F(x^i) \equiv 0$) and their types, limit cycles. It is especially useful in the studying of nonlinear differential equations because it does not require to solve them analytically. The dynamics of the system is described by determining the all different sets of initial conditions which leads to qualitatively different behavior, e.g. critical points, limit cycles or strange attractors.

Many results of this theory can obtained in analytical way. However, numerical simulations turn out to be a very powerful tool. It is especially useful in studying deterministic chaos. It is highly irregular behavior of trajectories (solutions) of a system, which seems to be stochastic but is generated by a deterministic mechanism.

One of the type of behavior exhibited in nonlinear dynamical systems is a limit cycle. It gives self-sustained oscillations with a constant period.

The bifurcation theory is a major part of research in dynamical economics. A central topic to this is the question whether the qualitative properties of a dynamical system change when one or more of the exogenous parameters are changing. The Hopf bifurcation leads from a critical point to a limit cycle. The bifurcation scenario is one of the ways to chaotic behavior.

Models described by systems of delay differential equations can be found in all areas of science. The functional differential equations are systems in which the current state depends not only on the present state but also on past or future states. Many economic phenomena can be modeled by systems with the feedback between present and past states. This way the past events influence the present evolution of systems. This history of the systems is required to be provided as an initial function.[1] The system of time delay differential equations is equivalent to the infinite dimensional system of ordinary differential equations. Such systems can exhibit cyclic behavior, for example a limit cycle after the Hopf bifurcation, or even chaotic dynamics (see [15]). The methods of this theory can be used to analyze economic models with different time lags.

Delay differential equations are such differential equations in which the derivatives of some unknown function at the present time $t$ depends on the value of the function at previous time $T$

$$\dot{x}(t) = F(t, x(t), g(x(t - T))), \tag{2}$$

where $g(x(t - T))$ is the feedback function, which depends on the state of the system at some time in the past $t - T$, for $T > 0$.

Since delay differential equations rely on the past history of the system in order to determine its current dynamics it is necessary to know the initial function as the initial condition instead of the initial point. The initial function gives us the states of the system in all past moments in interval $[t_0 - T, t_0]$.

In study of dynamics of DDEs as dynamical systems we need information about a state of equilibrium. They location does not depend on the value of time delay in the phase space because from definition they are stationary solutions. While the equilibria of DDEs are the same like the system without the delay, their stability changes significantly when the delay parameter $T$ is varied. The exponential stability is given by the spectrum of the linearization of DDEs in its equilibrium. This spectrum can be expressed as roots of an analytic function (a polynomial of exponential $\lambda \to \exp(-\lambda T)$) [30, 31].

## Business Cycle Model with Delay

### The Kaldor–Kalecki Model

The Kaldor–Kalecki model owes its origin to the Kaldor's concept of nonlinearity of investment function and Kalecki's concept of investment delay. Kaldor's idea of nonlinear model of business cycle [32] was given the following form by Chang and Smyth [16]

$$\dot{Y}(t) = \alpha[I(Y(t), K(t)) - S(Y(t), K(t))] \tag{3}$$

$$\dot{K}(t) = I(Y(t), K(t)) - \delta K(t), \tag{4}$$

[1] Using numerical methods to solve delay differential equations, a problem of derivative discontinuities is often present [29].

where $Y$ is income, $K$ is capital stock, $I(Y, K)$ is the investment function, $S(Y, K)$ is the saving function, $\delta$ is the capital depreciation rate and $\alpha$ is a parameter of speed of adjustment to an equilibrium. The important feature of this model is that the investment function $I(Y, K)$ is a nonlinear (s-shaped) function with respect to income $Y$.

Kalecki considered investment decisions depending on firm's savings and reasons to invest [33]. On the other hand, Kalecki assumed that there is an investment delay with an average time of investment realization in an economy $T$ [14]. First, investment orders $I$ are placed and new capital $D$ is delivered after time $T$.

$$D(t) = I(t - T). \tag{5}$$

The change of capital stock in time $t$ is connected with the delivery of new capital $D$ at time $t$ due to investment orders at time $t - T$. The capital accumulation equation (4) has now a new form

$$\dot{K}(t) = I(t - T) - \delta K(t). \tag{6}$$

Note that time lags are of different variety in economic processes. In the context of investment we can distinguish the delivery lags investment delay and time-to-build. The former is the situation when the investment is made at the beginning of investment period at time $t - T$ and productive capital is delivered at the end of this period at time $t$. In turn time-to-build denotes a period $T$ over which the investment are realized and capital goods are produced. The time delay in the Kaldor–Kalecki model is understood as the delivery lag [34, 35]. As Kalecki considered the average time for investment it can treated to be delivery lag not the time-to-build.

The depreciation term $\delta K(t)$ is taken in (6) at time $t$, and the $D(t) - \delta K(t)$ is net investment in time $t$. There are an alternative formulation especially used in the Solow model with delay, where the accumulation of capital stock is $\dot{K}(t) = I(K(t - T)) - \delta K(t - T)$ with no clear economic interpretation [36].

The new model of business cycle is formulated [19] where the Kaldor model is modified in such a way that the capital accumulation depends on past investment decisions. In this Kaldor–Kalecki model the cyclic behavior is due to the Kalecki time lag parameter and the limit cycle is created without the assumption of non-linearity on the investment and saving functions.

This model can be treated as the generalization of the Kaldor model in the following way. In the Kaldor model the cyclic behavior is represented by a limit cycle in the phase space. Whereas in the Kalecki theory basing on linear equations the cyclic behavior is represented by close trajectories (a center). As it is well known limit cycles are structurally stable whereas centers are destroyed by small perturbations of the right-hand sides of the dynamical systems.

It seems to be useful to construct the model of business cycle which possess a structurally stable limit cycle which is caused by the time delay in investment.

The dynamics of the Kaldor–Kalecki model is covered by the system of time-delay differential equations which is equivalent to an infinite dimensional autonomous dynamical system

$$\dot{Y}(t) = \alpha[I(Y(t), K(t)) - S(Y(t), K(t))] \tag{7}$$
$$\dot{K}(t) = I(Y(t-T), K(t)) - \delta K(t), \tag{8}$$

where investment $I(Y, K)$ and saving $S(Y, K)$ are functions of income $Y$ and capital stock $K$; the speed of adjustment in the market of goods is denoted by $\alpha$ and the capital depreciation rate by $\delta$. The parameter $T$ denotes time delay in the investment.

Consider some assumptions on the investment and saving functions such that the investment function separates on two functions of which the first depends only on income and the second on capital stock; the latter is linear in respect to its argument: $I(Y, K) = I(Y) + \beta K$, where $\beta < 0$. The linear saving function depends only on income: $S(Y) = \gamma Y$, where $\gamma > 0$. Taking into account these assumptions, system (7)–(8) can be rewritten to the form

$$\dot{Y}(t) = \alpha[I(Y(t)) + \beta K(t) - \gamma Y(t)] \tag{9}$$
$$\dot{K}(t) = I(Y(t-T)) + \beta K(t) - \delta K(t). \tag{10}$$

Let us consider system (9)–(10) as a second order time delay differential equation

$$\ddot{Y}(t) + f(Y(t))\dot{Y}(t) + g(Y(t)) = 0, \tag{11}$$

where

$$f(Y) = -\alpha(I_Y - \gamma) - (\beta - \delta),$$
$$g(Y) = -\alpha\gamma(\beta - \delta)Y(t) + \alpha(\beta - \delta)I(Y) - \alpha\beta I(Y(t-T)).$$

We linearize equation (11) around a fixed point $(Y^*, K^*)$. We define a new variable as the deviation from the critical point $y = Y - Y^*$ and then the characteristic equation has the form

$$\lambda^2 + A\lambda + B + De^{-\lambda T} = 0, \tag{12}$$

where

$$A = -\alpha(i_y(0) - \gamma) - (\beta - \delta)$$
$$B = -\alpha\gamma(\beta - \delta) + \alpha(\beta - \delta)i_y(0)$$
$$D = -\alpha\beta i_y(0).$$

The characteristic equation (12) is of the transcendental type. Such an equation has infinite number of eigenvalues which usually cannot be found analytically. The former is a consequence that the time delay differential equations are equivalent to an indefinite set of ordinary differential equations.

To determine the existence of a cyclic solution in the Kaldor–Kalecki model we use the Poincaré–Andronov–Hopf bifurcation theorem (also called the Hopf bifurcation) [9, p. 151–152]. This kind of bifurcation causes the change of stability of the critical point and the creation of a limit cycle around it. In our case a critical point (a stable focus) looses it stability and a stable limit cycle is a result. Let there is no real eigenvalues and complex conjugate eigenvalues exists $\lambda = \sigma \pm i\omega$. There are two conditions of the limit cycle creation (a) a real part $\sigma$ of a pair of complex conjugate eigenvalues (with nonzero imaginary part) changes a sign from negative to positive as a value of a bifurcation parameter increases, and (b) the derivative of the real part of eigenvalue with respect to a bifurcation parameter is positive as a real part $\sigma$ passes through zero.

We consider the time delay parameter $T$ as the bifurcation parameter. Assuming that eigenvalues have a form $\lambda = \sigma + i\omega$, we rewrite (12) as a system of its real and imaginary parts

$$\sigma^2 - \omega^2 + \sigma A + B + De^{-\sigma T} \cos \omega T = 0 \tag{13}$$

$$2\sigma\omega + \omega A - De^{-\sigma T} \sin \omega T = 0. \tag{14}$$

The first condition of the Hopf bifurcation theorem is that bifurcation occurs when the complex conjugate pair of eigenvalues is purely imaginary, i.e., $sigma = 0$. In this case the system of eigenvalue equations (13)–(14) reduces to the form

$$-\omega^2 + B + D \cos \omega T = 0 \tag{15}$$

$$\omega A - D \sin \omega T = 0. \tag{16}$$

Solving it for $\omega$ and $T$ can see that there is an infinite number of solutions

$$\omega T = \omega_{\text{bi}} T_{\text{bi}} + 2\pi n, \qquad n = 1, 2, \ldots .$$

Therefore, it is more convenient to use a new parametric variable $v = \omega T$, which is the argument of the trigonometric functions in system (15)–(16). From the second equation of system (15)–(16) the solution for $\omega$ is found in the form

$$\omega = \frac{D \sin v}{A}. \tag{17}$$

The result (17) is, in turn, substituted into the first equation of (15)–(16), and with the help of the trigonometric identity $\sin^2 v = 1 - \cos^2 v$ we obtain

$$D^2 \cos^2 v + DA^2 \cos v + BA^2 - D^2 = 0. \tag{18}$$

This equation gives us the solution for $v$, which used in (17) gives us the solution for $\omega$ and finally

$$T_{\text{bi}} = \frac{v}{\omega} = \frac{A}{D} \frac{v}{\sin v} = \frac{A}{D} \frac{1}{\frac{\sin v}{v}}, \tag{19}$$

The bifurcation value of the bifurcation parameter $T_{\text{bi}}$ depends on the rest parameters of the model

$$T_{\text{bi}} = \frac{A}{D}\frac{v}{\sin v} = \frac{A}{D}\frac{\arccos z}{\sin(\arccos z)} = \frac{A}{D}\frac{\arccos z}{\frac{\pi}{2} - z}, \qquad (20)$$

where

$$z_{1,2} = \frac{1}{2D}\left(-A^2 \mp \sqrt{A^4 - 4(A^2B - D^2)}\right).$$

The next step is to check the transversality condition. Let us differentiate the eigenvalue equation (12)

$$\text{sign}\left[\frac{\partial \sigma}{\partial T}|_{T=T_{\text{bi}}}\right] = \text{sign}\left[D(A\omega \sin \omega T + 2\omega^2 \cos \omega T)\right]. \qquad (21)$$

After substituting equation (17) to (21) we obtain

$$\text{sign}\left[\frac{\partial \sigma}{\partial T}|_{T=T_{\text{bi}}}\right] = \text{sign}\left[\frac{A^2}{2} + \cos \omega T\right],$$

that the transversality condition is satisfied for any delay parameter $T$ provided that

$$\frac{A^2}{2} + \cos \omega T > 0.$$

Additionaly, we can determine the period of a cycle

$$P = \frac{2\pi}{|\omega(T_{\text{bi}})|} = \frac{2\pi T_{\text{bi}}}{|v|}$$

and the radius of an orbit by $R \propto \sqrt{T}$ [21].

## The Kaldor–Kalecki Model in Small Time Lag Approximation

Equation (11) is a functional differential equation. Such equations are a generalization of the ordinary differential equation by allowing the state of the system at time $t$ to depend on past history of the system.

For the further analysis it is useful to investigate properties of the system in the linear approximation of the retarded investment function with respect to lag argument $T$. Such a form of the model will called the first order approximation of the Kaldor–Kalecki model. For this end it is convenient to represent $g(Y)$ in the following form

$$g(Y(t)) = \alpha\beta T\dot{Y}(t)\frac{I(Y(t-T)) - I(Y(t))}{T\dot{Y}(t)}$$
$$- \alpha\delta I(Y(t)) - \alpha\gamma(\beta-\delta)Y(t)$$
$$= -\alpha\delta I(Y(t)) - \alpha\gamma(\beta-\delta)Y(t) + \Delta g(Y(t)) \tag{22}$$

and

$$\Delta g(Y(t)) \overset{T\to 0}{=} \alpha\beta I\frac{\mathrm{d}I}{\mathrm{d}t}. \tag{23}$$

To make the approximation for small $T$ it is taken only the linear terms of expansion of the retarded income function $Y(t-T)$ with respect to $T$ such that $Y(t-T) = T(t) - T\dot{Y}(t) + \mathcal{O}(T^2)$ and than neglected all terms but linear ones in the expansion of investment function $I(Y(t-T)$ [37]

$$I(Y(t-T)) = I(Y - T\dot{Y} + \cdots) = I(Y) - T\dot{Y}I_Y + \cdots,$$

Therefore, if income function $Y(t)$ is a slowly changing function of time $t$, the infinite dimensional dynamical system is reduced to a two-dimensional one for $T \ll 1$. Unfortunately some nice properties of the infinite dimensional K-K model cannot be inherited by the new system, namely the generation of limit cycles without the Kaldor condition of non-linearity of investment function.

Now it allows us to formulate the following corollary

**Corollary 1** *The Kaldor–Kalecki model in the first order approximation can be reduced to the form of the Liénard equation*

$$\ddot{Y} + \alpha\left(-\gamma\frac{\beta}{\alpha} + \frac{\delta}{\alpha}\beta T I_Y(Y) - I_Y(Y)\right)\dot{Y}$$
$$- \alpha\delta I(Y) - \alpha\gamma(\beta-\delta)Y = 0. \tag{24}$$

If we put $x = Y$ and $\dot{y} = -g(x)$ then (24) can be written as the Liénard system

$$\dot{x} = y - \mathcal{F}(x), \tag{25}$$
$$\dot{y} = -g(y), \tag{26}$$

where

$$g(x) = \alpha\delta I(x) - \alpha\gamma(\beta-\delta)x$$

$$\mathcal{F}(x) = \int_0^x f(u)du,$$

$$f(u) = -\alpha I_u + A + \alpha\beta T I_u, \quad A = \alpha\gamma - \beta + \delta > 0.$$

Formally, assuming $T = 0$ one can obtain the Kaldor model in the Liénard representation from which after the substitution

$$f(Y) \to f(Y) + \alpha\beta T I_Y \mathcal{F}(x) \to \mathcal{F}(x) - \alpha|\beta|T I(x)$$

the Kaldor–Kalecki model in the first order approximation can be obtained.

Let the investment function $I(x)$ has the $s$-shaped property, i.e. $I_{xx} > 0$ for $x < x_0$ and $I_{xx} < 0$ for $x > x_0$ and $x_0$ is the inflexion point. By expanding this function in a neighborhood of the inflexion point one can obtain the investment function $I(x)$ truncated on third order terms

$$\begin{aligned} I(x) = I(x_0) &+ \frac{dI}{dx}|_{x=x_0}(x - x_0) \\ &+ \frac{1}{6}\frac{\mathrm{d}^3 I}{\mathrm{d}x^3}|_{x=x_0}(x - x_0)^3 + \cdots, \end{aligned}$$

where all even derivatives of $I$ in $x_0$ are zero because the function $I$ is odd in $(x - x_0)$. Such an approximated function is $s$-shaped what is the Kaldor condition.

To simplify the analysis one can put the depreciation rate $\delta = 0$. After introducing a new variable $z \equiv x - x_0$, system (25)–(26) becomes

$$\dot{z} = y - a_1 z - a_3 z^3 - a_0 \tag{27}$$

$$\dot{y} = -\kappa^2 z - s, \tag{28}$$

where

$$\begin{aligned} a_0 &= -\alpha(1 + |\beta|T)I(0) + (\alpha\gamma - \beta)x_0 \\ a_1 &= -(\alpha + |\beta|T)I_z(0) - (\beta - \alpha\gamma) \\ a_3 &= -\frac{1}{6}\alpha(1 + |\beta|T)I_{zzz}(0) \\ s &= \kappa^2 x_0 > 0 \end{aligned}$$

In (27)–(28) the value of $x_0$ can be chosen in such a way that $a_0$ equals zero. One can rescale the variables in system (27)–(28) to distinguish parameters crucial for the dynamics and construct the dimensionless model

$$y \to \bar{y} = ay \tag{29}$$

$$z \to \bar{z} = bz \tag{30}$$

$$t \to \bar{y} = ct \tag{31}$$

$$s \to \bar{s} = \frac{a}{c}s. \tag{32}$$

Such a transformation preserve the form of the system if the condition $\frac{a}{bc} = \frac{1}{\kappa^2}$ is satisfied. Then the system (27)–(28) takes the form

$$\frac{d\bar{z}}{d\bar{t}} = \frac{\kappa^2}{c^2}\bar{y} - \frac{\kappa^2 a}{c^2}\left[\alpha(1+|\beta|T)\left(I_{\bar{z}}(0)\bar{z} + \frac{1}{6}I_{\bar{z}\bar{z}\bar{z}}(0)\bar{z}^3\right)\right] - (\alpha\gamma - \beta)\bar{z} \tag{33}$$

$$\frac{d\bar{y}}{d\bar{t}} = -(\bar{z} + \bar{s}), \tag{34}$$

Another approach is based on a direct application of the Hopf theorem. Let choose $c = \kappa$ in (33)–(34). If one additionally assume that $a = \alpha^{-1}$ and take the expansion of $I(x)$ to $(2n+1)$-order then one obtain the following system

$$\frac{\mathrm{d}z}{\mathrm{d}t} = y - a_1 z - a_3 z^3 - \cdots - a_{2n+1} z^{2n+1} \tag{35}$$

$$\frac{\mathrm{d}y}{\mathrm{d}t} = -z - s, \tag{36}$$

where

$$a_1 = -(1+|\beta|T)I_z(0) + A$$

$$a_3 = -\frac{1}{6}(1+|\beta|T)I_{zzz}(0)$$

$$\vdots$$

$$a_{2n+1} = -\frac{1}{(2n+1)!}(1+|\beta|T)\frac{\mathrm{d}^{2n+1}I(0)}{\mathrm{d}z^{2n+1}}.$$

Let us assume, now, that all the parameter, also $s$, are constant, with the exception of the delay parameter $T$. One can show that there exists a bifurcation value of the parameter $T$. The right-hand sides of the system (35)–(36) define the one-parameter family of vector fields on $\mathbb{R}^2$.

$$X^1(z, y) = y - a_1 z - a_3 z^3 - \cdots - a_{2n+1} z^{2n+1} \tag{37}$$

$$X^2(z, y) = -z - s. \tag{38}$$

In contrast to the previous case, let consider the case when $s = 0$ and $n = 1$. Then the vector field $X_{a_i}(0, 0) = 0$ and for any $a_i$ the tangent vector takes the form

$$dX_{a_i}(0, 0) = \begin{pmatrix} -\alpha\gamma + \beta + (1+|\beta|T)I_z(0) & 1 \\ -1 & 0 \end{pmatrix}. \tag{39}$$

Eigenvalues of linearization matrix can be written by the help of parameter $\bar{a}$ where

$$\bar{a} = -\alpha\gamma + \beta + (1+|\beta|T)I_z(0). \tag{40}$$

We consider only such values of parameters $\alpha, \beta, \gamma, I_z(0), T$, for which $\bar{a}^2 < 4$, i.e. eigenvalues are imaginary ($\Im\lambda(\bar{a}) \neq 0$) and

$$\lambda(\bar{a}) = \frac{1}{2}\left(\bar{a} \pm \mathrm{i}\sqrt{4-\bar{a}^2}\right). \tag{41}$$

Therefore

$$\begin{aligned} -2 < \bar{a} < 0 \quad & \operatorname{Re}\lambda(\bar{a}) < 0 \\ \bar{a} = 0 \quad & \operatorname{Re}\lambda(\bar{a}) = 0 \\ 0 < \bar{a} < 2 \quad & \operatorname{Re}\lambda(\bar{a}) > 0 \end{aligned}$$

and

$$\frac{\mathrm{d}}{\mathrm{d}\bar{a}}(\operatorname{Re}\lambda(\bar{a})|_{\bar{a}=0} = \frac{1}{2}. \tag{42}$$

Now one can apply the Hopf theorem which says that for one-parameter family of closed orbits for the vector field $X = (X_{\bar{a}}, 0) = (X_{a_1}, 0)$ in the neighborhood of point $(0, 0, 0)$. Let us note that truncating the expansion of investment function on the third order terms is not substantial and the analysis is valid for any $n$ because the first term of expansion is essential. The period of closed orbit is well approximated by $T \simeq 2\pi/|\lambda(0)|$, while amplitude of oscillations is proportional to $\sqrt{\bar{a}}$. The presence of a new term with $T$ in $a$ causes the amplitude to be greater with $\Delta A$ for small modifications

$$\begin{aligned} A = \sqrt{\bar{a}} &= \sqrt{a_{\text{Kaldor}} + (1+|\beta|T)I_z(0)} \\ &\cong \sqrt{a_{\text{Kaldor}}}\left(\sqrt{1 + \frac{1}{2}\frac{(1+|\beta|T)I_z(0)}{a_{\text{Kaldor}}}}\right) \\ &= A_{\text{Kaldor}} + \Delta A. \end{aligned} \tag{43}$$

The bifurcation value $T = T_{\text{bi}}$ is determined by

$$T_{\text{bi}} = \frac{1}{|\beta|}\left(\frac{A}{I_z(0)} - 1\right). \tag{44}$$

One can see that the bifurcation always exists when $A > I_z(0)$.

**Corollary 2** *In the Kaldor–Kalecki model there is a bifurcation to a limit cycle via the Hopf bifurcation mechanism.*

***Proof*** Let assume that all parameters but $T$ are constant and $\lambda = \lambda(a(T))$. Then

$$\frac{\mathrm{d}\operatorname{Re}\lambda(a(T))}{\mathrm{d}T} = \frac{\mathrm{d}\operatorname{Re}\lambda(a(T))}{\mathrm{d}a}\frac{\mathrm{d}a}{\mathrm{d}T} = \frac{1}{2}|\beta|I_z(0) > 0.$$

Therefore the assumptions of the Hopf theorem are fulfilled if $I_z(0) > 0$, and $T = T_{\text{bi}}$ is a bifurcation parameter.

The condition of stability or instability of the limit cycle was determined on the base of the Guckenheimer stability parameter method [9].

In this example the Hopf bifurcation theory is applied to analyze the fragility of dynamics of the Kaldor–Kalecki model with respect to the time delay parameter. The time-to-build parameter plays role the bifurcation parameter. From this analysis the following conclusions important in control of business cycle can be given

- The Kaldor–Kalecki model can be represented by the Liénard system for the small time delay parameter $T$.
- For small parameter $T$ there is a bifurcation value of this parameter. A stable limit cycle is generated via the Hopf bifurcation scenario when the investment function is $s$-shaped.
- In general case of the Kaldor–Kalecki model (i.e. for any value of parameter $T$) there is a mechanism of creating the close orbit even for the linear investment function.
- In general, the time delay parameter $T$ makes the amplitude of the cycle increasing. In general, the period of oscillations does not depend on a lag parameter.

## The Kaldor–Kalecki Model as a System with Self-sustaining Oscillations

Let us present the classification of vibrating systems. We have a nonlinear autonomous dynamical system with one degree of freedom

$$\ddot{x} + S(x) = Q(x, \dot{x}), \tag{45}$$

where $x$ is a positional variable, a dot means the differentiation with respect to time $t$. Equation (45) can be obtained from

$$\frac{\mathrm{d}}{\mathrm{d}t}\left(\frac{1}{2}\dot{x}^2\right) + S(x)\dot{x} = Q(x, \dot{x})\dot{x} \tag{46}$$

or

$$\frac{\mathrm{d}}{\mathrm{d}t}(E + V(x)) = \dot{x}Q(x, \dot{x}) = vQ(x, v), \tag{47}$$

where $E$ is the kinetic energy and $V(x) = \int_0^x S(\eta)d\eta$ is the potential energy.

The change of energy of the system $E + V$ is forced by a non-conservative force—$vQ(x, v)$. Depending on the sign of this contribution, energy is added or substrate from the system and we have three possibilities

(A) if $vQ(v, x) > 0$ for $v \neq 0$ on whole phase plane $(x, v)$, then the system gains energy along the phase curve;

(B) if $vQ(v, x) < 0$ for $v \neq 0$ on whole phase plane $(x, v)$, then the system dissipates energy along the phase curve—*dissipative system*;
(C) If in some regions of the phase space $(x, v)$ hold $vQ(v, x) > 0$, and in other regions $vQ(x, v) < 0$, then the system is called *self-sustaining*.

Note that the system is conservative when the non-conservative force $vQ(x, v)$ is absent, i.e., if $vQ(v, x) = 0$ for $v \neq 0$ on whole phase plane $(x, v)$, and the system moves along the phase curve with energy conserved.

First, consider the classification of vibration against the source

- free vibrations—takes place in the autonomous systems without the action of outside energy source or outside forces—they are generated by initial conditions;
- forced vibrations—hold in non-autonomous dynamical systems with time-dependent force;
- parametric vibrations—hold in non-stationary systems and are caused by the change of model parameters;
- self-sustained vibrations—hold in non-conservative systems, where inflow of energy to the system is regulated by vibrations and is originated from the non-vibrating system.

Second, consider the classification of vibrations against the solution

- periodic oscillations;
- quasi-periodic oscillation—there are at least two frequency with irrational ratio;
- chaotic oscillations—appear in the nonlinear systems either non-autonomous with one degree of freedom or autonomous with at least three dimensions.

Self-sustaining vibrations are related with the energy transfer between the source and the system and vice versa. For any vibrating system which interact with surroundings the amplitude of these vibrations is damped over the time. If the loss of energy is compensated by other source of energy depending on the state of system, then the equilibrium is established between the lost and delivered energy. The vibrates periodically, and its move is represented in the phase space by a limit cycle.

Self-sustaining systems are

- the systems which have the source of energy, which does not generate vibrations;
- the systems which is a source of vibration itself;

Physical examples are electric bell, violin sound, squeaking of hinges.

In the case of the Kaldor–Kalecki model with the small delivery lags presented in the previous section we have

$$\ddot{Y} + \alpha\delta I(Y) - \alpha\gamma(\beta - \delta)Y = \alpha\left(\gamma\frac{\beta}{\alpha} - \frac{\delta}{\alpha}\beta T I_Y(Y) + I_Y(Y)\right)\dot{Y} \tag{48}$$

or

$$\ddot{Y} + S(Y) = Q(Y, \dot{Y}) \tag{49}$$

where

$$S(Y) = -\alpha\gamma(\beta - \delta)Y \tag{50}$$

$$Q(Y, \dot{Y}) = \alpha\left(\gamma\frac{\beta}{\alpha} - \frac{\delta}{\alpha}\beta T I_Y(Y) - I_Y(Y)\right)\dot{Y}. \tag{51}$$

Let us rewritten (51) to the form

$$Q(Y, \dot{Y}) = v\dot{Y} = \alpha\left[\gamma\frac{\beta}{\alpha} + \left(1 - \delta T\frac{\beta}{\alpha}\right) I_Y(Y)\right]\dot{Y}. \tag{52}$$

Because the parameter $\beta$ is negative, then $v$ is always positive. In previous section it was shown that the Kaldor–Kalecki model possesses a limit cycle solution, and variable $Y$ increases and decreases cyclically. Therefore, the $\dot{Y}$ is positive and negative respectively over a cycle. Similarly, $Q(Y, \dot{Y})$ is positive and negative in some regions of the phase space and the cycle is self-sustaining.

## The Kaldor–Kalecki Growth Model

In 1984 Dana and Malgrange introduced the exponential growth of the autonomous demand $G_0e^{gt}$ to the Kaldor model of business cycle [38]. In a similar way we modified the Kaldor–Kalecki business cycle model and obtain the growth model in the form [20]

$$\dot{Y}(t) = \alpha[I(Y(t), K(t)) - S(Y(t), K(t)) + G_0e^{gt}] \tag{53}$$

$$\dot{K}(t) = I(Y(t - T), K(t)) - \delta K(t), \tag{54}$$

where $G_0$ and $g$ are constants.

We assume, without loss of the generality, that a linear saving function $S$ depends on income $Y$ only

$$S(Y, K) = \gamma Y \tag{55}$$

and an s-shaped investment function $I$ has the form [38]

$$I(Y, K) = K\Phi\left(\frac{Y}{K}\right) = K\Phi(x) \tag{56}$$

with

$$\Phi(x) = c + \frac{\mathrm{d}}{1 + \exp[-a(vx - 1)]}, \tag{57}$$

where $x = Y/K$ denotes income per capital stock and $\Phi(x)$ satisfies Kaldor's conditions: $\Phi_{xx}(x) > 0$ for $x < x^*$, $\Phi_{xx}(x^*) = 0$, and $\Phi_{xx}(x) < 0$ for $x > x^*$.

We introduce new variables to obtain the stationary solution of system (53)–(54) in the form of constant rate growth path

$$k = Ke^{-gt}, \quad y = Ye^{-gt}. \tag{58}$$

Now taking above assumptions, the model takes the form

$$\dot{y}(t) = \alpha[k(t)\Phi(y(t)/k(t)) - \gamma y(t) + G_0] - gy(t), \tag{59}$$

$$\dot{k}(t) = k(t)\Phi(y(t-T)/k(t)) - (g+\delta)k(t) \tag{60}$$

and $\Phi(y/k) = \Phi(Y/K) = \Phi(x)$.

The system (59)–(60) has a unique fixed point $(y^*, k^*)$ with positive coordinates

$$y^*(\alpha) = x^* k^*(\alpha) \tag{61}$$

$$k^*(\alpha) = \frac{\alpha G_0}{gx^* + \alpha(sx^* - (g+\delta))}, \tag{62}$$

where $x^*$ is the unique solution of the equation

$$\Phi(x^*) = g + \delta.$$

For this s-shape function $\Phi(x)$, the critical point $x^*$ always exists provided that $c < g + \delta < c + d$.

Let linearize the system at the critical point $(y^*, k^*)$

$$\frac{\mathrm{d}}{\mathrm{d}t}(y - y^*) = [\alpha(I_y(y^*) - \gamma) - g](y - y^*) + \alpha I_k(k^*)(k - k^*) \tag{63}$$

$$\frac{\mathrm{d}}{\mathrm{d}t}(k - k^*) = I_y(y^*)(y - y^*)(t - T) + [I_k(k^*) - (g+\delta)](k - k^*) \tag{64}$$

and use the transformation $\bar{y} = y - y^*$, $\bar{k} = k - k^*$

$$\dot{\bar{y}} = [\alpha(I_{\bar{y}}(0) - \gamma) - g]\bar{y}(t) + \alpha I_{\bar{k}}(0)\bar{k}(t)$$

$$\dot{\bar{k}} = I_{\bar{y}}(0)\bar{y}(t - T) + [I_{\bar{k}}(0) - g - \delta]\bar{k}(t).$$

Let calculate the Jacobian $J$ and then the characteristic equation for the critical point $y^* = 0$, $k^* = 0$ (henceforth the bar over variables is deleted)

$$\lambda^2 + b\lambda + a = de^{-\lambda T}, \tag{66}$$

where

$$b = -[\alpha(I_y(0) - \gamma) + I_k(0) - 2g - \delta]$$
$$a = [\alpha(I_y(0) - \gamma) - g][I_k(0) - g - \delta]$$
$$d = \alpha I_y(0) I_k(0) < 0.$$

Equation (66) is a transcendental equation which possesses an infinite number of solutions.

As in the previous section we use the Hopf bifurcation theorem [9, p. 151-152] to determine the existence of the cyclic solution. Again, we consider the time delay parameter $T$ as the bifurcation parameter. All the rest parameters are assumed to be constant.

The amplitude of this cycle is proportional to the value of the trace of the Jacobian tr $J$. Because tr $J$ is a coefficient of the linear term in (66), then tr $J = 0$ is equivalent to consider $e^{-\lambda T} \cong 1 - \lambda T$ and

$$T = T_{\text{bi}} = \frac{b}{d} = -\frac{\alpha(I_y(0) - \gamma) + I_k(0) - 2g - \delta}{\alpha I_y(0) I_k(0)}. \tag{67}$$

We recall that the conditions for investment function are $I_y(0) > 0$, $I_k(0) < 0$. Therefore, the bifurcation value of the bifurcation parameter $T_{\text{bi}}$ is positive provided that

$$\alpha(I_y(0) - \gamma) + I_k(0) > 2g + \delta. \tag{68}$$

Let us turn to the analysis of the stability of the critical point $(y^*, k^*)$. The trivial critical point of system (59)–(60) is stable if all roots of the characteristic equation (66) have negative real parts. It becomes unstable when a sign of a real part of an eigenvalue $\lambda$ is positive. Let us consider two cases.

In the first case, an eigenvalue is real and it changes a sign. It takes place when $d = a$, which requires $a < 0$ ($\alpha(I_y(0) - \gamma) - g > 0$), because $I_k(0) \equiv \beta < 0$. The position feedback loop has a positive feedback.

In the second case, a pair of complex eigenvalues crosses the imaginary axis ($\lambda = \pm i\omega, \omega > 0$). In our model it happens when the following equations are satisfied

$$a - \omega^2 = d \cos \omega T \tag{69}$$
$$b\omega = -d \sin \omega T. \tag{70}$$

From system (69)–(70) we obtain a quartic polynomial after adding the squared sides of this system. The condition $b^2 - 2a > 0$ is always satisfied for any parameters of the model, therefore the unique solution of this polynomial is

$$\omega_+ = \sqrt{\frac{2a - b^2}{2} + \sqrt{\left(\frac{2a - b^2}{2}\right)^2 + d^2 - a^2}}. \tag{71}$$

The solution $\omega_+$ is real and positive, provided that both the condition $b^2 - 2a > 0$ and the condition $d^2 - a^2 > 0$ are fulfilled.

Finally we solve system (69)–(70) for the bifurcation parameter $T$ using the found $\omega_+$. We rewrite equations of system (69)–(70)

$$\cos\omega_+ T = \frac{a - \omega_+^2}{d} \tag{72}$$

$$\sin\omega_+ T = -\frac{b\omega}{d} \tag{73}$$

and obtain the solution for the delay parameter $T$ as a function of the other system parameters

$$T_j = \begin{cases} \frac{1}{\omega_+}\left[2\pi j + \arccos\left(\frac{a-\omega_+^2}{d}\right)\right], & b > 0 \\ \frac{1}{\omega_+}\left[2\pi(j+1) - \arccos\left(\frac{a+\omega_+^2}{d}\right)\right], & b < 0, \end{cases} \tag{74}$$

where $j = 0, 1, 2, \ldots$, and the argument of an arc-cosine function is the interval $[0, \pi]$.

The next step is to check the transversality condition. It means that eigenvalues of the characteristic equation must cross the imaginary axis with the positive speed. After differentiation of the characteristic equation over $T$ and substitution equation (66) we obtain

$$\frac{\partial\lambda}{\partial T} = -\frac{\lambda(\lambda^2 + b\lambda + a)}{2\lambda + b}. \tag{75}$$

Then after substitution $\lambda = \sigma + i\omega$ and using (69) and (71) we obtain that the transversality condition is always positive

$$\frac{\partial}{\partial T}(\mathrm{Re}\lambda)\bigg|_{T=T_{\mathrm{bi}}} = \frac{\partial\sigma}{\partial T}\bigg|_{T=T_{\mathrm{bi}}} > 0. \tag{76}$$

As both conditions of the Hopf bifurcation theorem are fulfilled we can state that there is a limit cycle solution in the Kaldor–Kalecki growth model. Let us note that because (66) has an infinite number of roots, the steady state is mostly a saddle point.

Here, to determine whether this Hopf bifurcation lead to a stable or unstable limit cycle, i.e., whether the Hopf bifurcation is supercritical or subcritical, let us consider a numerical approach.

To this aim we consider the investment function (57). Dana and Malgrange estimated it using the French quarterly data for 1960–1974 and obtained

$$\Phi(x) = 0.01 + \frac{0.026}{1 + \exp[-9(4.23x - 1)]}. \tag{77}$$

Additionally we assume that $\alpha = 0.7$, $\gamma = 0.15$ and $\delta = 0.007$. Then the Hopf bifurcation takes place for $T_{\text{bi}} \approx 1$. Figure 6.1 shows the focus for $T_{\text{bi}} > T = 0.5$ and the limit cycle $T_{\text{bi}} < T = 2$. The limit cycle is stable.

## Conclusion

In physics as well as in economics we are looking for dynamical models as simple as possible following Einstein's desiderata. Then, such elaborated models are extended, generalized for the description of a degree of complexity which should be adequate to describe the reality.

In the paper we demonstrated that delay differential equation are effective in description degree of complexity of trade cycle or economic growth models.

In the presented models we are looking for cyclic and complex behavior of trajectories in the phase space. In this investigation we can find some strict analogy to the models which work well in physics like the Mackey-Glass model, model of self-sustaining cycles in nonlinear Lienard type dynamical systems and many other models.

Michael Atiyah and Gregory W. Moore speculate about the role of relativistic versions of delayed differential equations in fundamental physics. Authors argue relativistic invariance implies that we must consider both advanced and retarded terms in the equations, so we refer to them as shifted equations [39].

It is interesting that some stochastic laws in physics can be derived from the concept of delay differential equation itself. It is a well know fact that deterministic Brownian motion can be generated from differential delay equations [40]. Therefore the source the complexity can lie in endogenous models with delay itself.

The first step in looking for complexity in economic model is study cyclic solutions in delay differential equation. They can emerge principally in two ways: either in non-autonomous systems (i.e., parametric resonance) or via Hopf bifurcation in autonomous one. We concentrate our attention on the last case and apply local Hopf bifurcation methods. The Hopf cycle appears when a local fixed point loses or gains stability and cycle appears due to a change in a time to build/delivery time parameter. For value of $T = T_{\text{bi}}$ which is corresponding bifurcation value the Hopf bifurcation takes place there is a corresponding initial function $\phi(t)$; $t$ belongs to interval$(-T_{\text{bi}}, 0)$ that limits dynamics to lie on the center manifold (spanned on the eigenvectors corresponding to purely imaginary roots of characteristic equation). For value $T$ different from $T_{\text{bi}}$ steady state form in the generic case a saddle because there are an infinite number of complex roots with the positive and negative parts. Finally we obtain Hopf saddle cycles as a generic solution in the future. Just this solution is interpreted as a trade cycle.

We shown that the dynamic effects in the Kaldor–Kalecki model has started even with a short time lag. This nonlinear dynamic effects manifests through emerging a limit cycle.

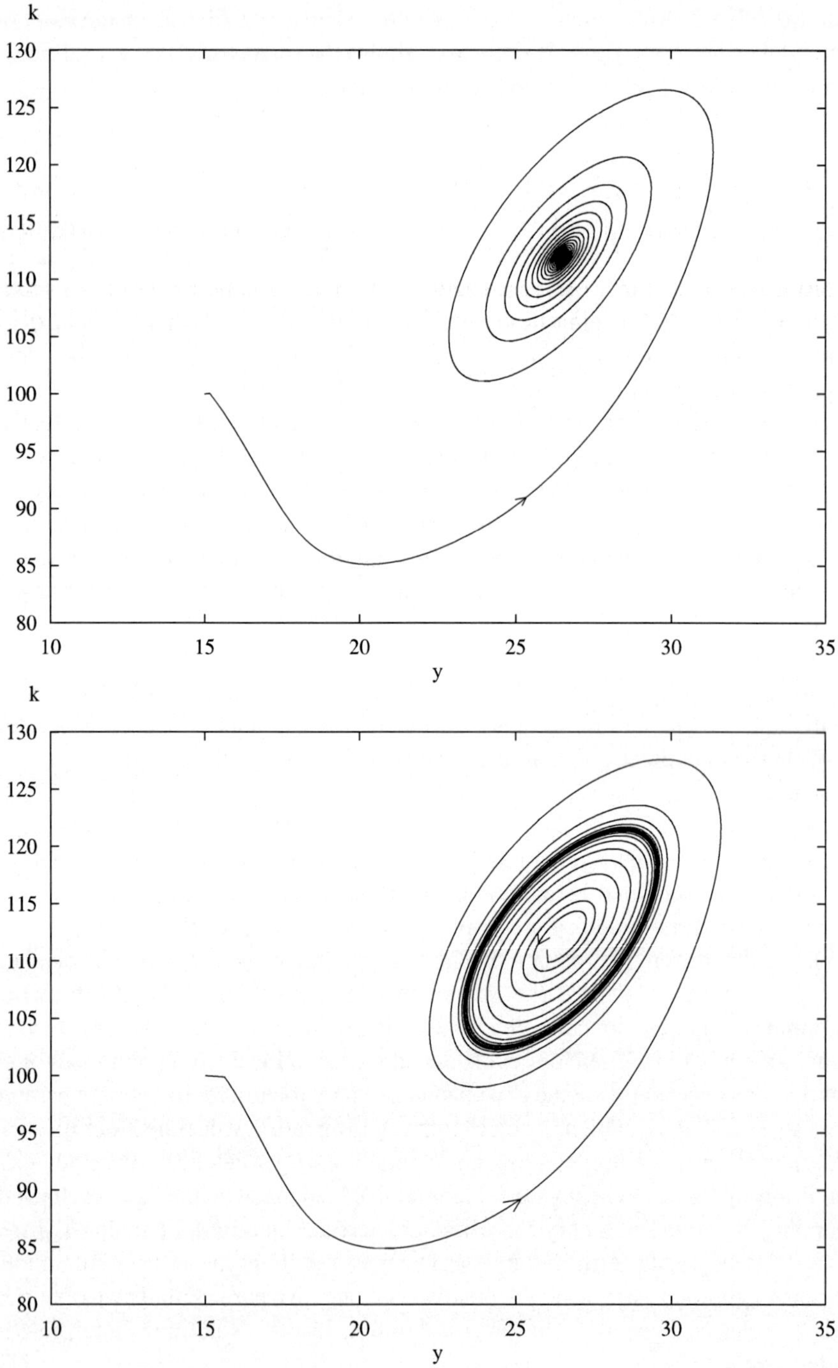


**Fig. 6.1** Phase portraits for the Kaldor–Kalecki growth model; upper panel: the stable focus $T = 0.5$, lower panel: the stable limit cycle $T = 2$

**Acknowledgements** The authors acknowledge the support of the Narodowe Centrum Nauki (National Science Centre Poland) project 2014/15/B/HS4/04264. We are grateful to prof. Marcel Ausloos for comments and useful suggestions.

## References

1. F.M. Atay (ed.), *Complex Time-Delay Systems* (Springer, Berlin, 2010)
2. W.B. Arthur, Complexity and the economy, *The Complexity Vision and the Teaching of Economics*, ed. by D. Colander, Ch. 1 (Edward Elgar, Cheltenham, 2000), pp. 93–110
3. M. Sportelli, M. Mastrorillo, Non-linear dynamics and complexity in economics: a historical overview (2014). https://doi.org/10.2139/ssrn.2548210
4. L. Piscitelli, M.C. Sportelli, A simple growth-cycle model displaying "Silnikov chaos", in *Economic Complexity*, eds. by W.A. Barnett, C. Deissenberg, G. Feichtinger. Non-linear Dynamics (Multi-agents Economies and Learning, Emerald, Bingley, 2004), pp. 3–30
5. C. Chiarella, *The Elements of a Non-linear Theory of Economic Dynamics* (Springer, Berlin, 1990)
6. C. Chiarella, P. Flaschel, Real and monetary cycles in models of Keynes-Wicksell type. J. Econ. Behav. Organ. **30**(3), 327–351 (1996). https://doi.org/10.1016/S0167-2681(96)00873-6
7. C. Chiarella, P. Flaschel, *The Dynamics of Keynesian Monetary Growth* (Cambridge University Press, Cambridge, 2000)
8. M. Boldrin, L. Montrucchio, On the indeterminacy of capital accumulation paths. J. Econ. Theory **40**(1), 26–39 (1986). https://doi.org/10.1016/0022-0531(86)90005-0
9. J. Guckenheimer, P. Holmes, *Nonlinear Oscillations, Dynamical Systems, and Bifurcations of Vector Fields* (Springer, New York, 1983)
10. K. Nishimura, M. Yano, Nonlinear dynamics and chaos in optimal growth: an example. Econometrica **63**(4), 981–1001 (1995). https://doi.org/10.2307/2171805
11. A. Medio, Continuous-time models of chaos in economics. J. Econ. Behav. Organ. **16**, 115–151 (1991)
12. M. Sportelli, A Kolmogoroff generalized predator-prey model of Goodwin's growth cycle. J. Econ. **61**(1), 35–64 (1995). https://doi.org/10.1007/BF01231483
13. P. Asea, P. Zak, Time-to-build and cycles. J. Econ. Dyn. Control **23**(8), 1155–1175 (1999). https://doi.org/10.1016/S0165-1889(98)00052-9
14. M. Kalecki, A macrodynamic theory of business cycles. Econometrica **3**(3), 327–344 (1935)
15. J.K. Hale, S.M. Verduyn Lunel, *Introduction to Functional Differential Equations*. Applied Mathematical Sciences, vol. 99 (Springer, New York, 1993). https://doi.org/10.1007/978-1-4612-4342-7
16. W. Chang, D. Smyth, The existence and persistence of cycles in a non-linear model: Kaldor's 1940 model re-examined. Rev. Econ. Stud. **38**(1), 37–44 (1971). https://doi.org/10.2307/2296620
17. S. Ichimura, Toward a general nonlinear macrodynamics theory of economic fluctuations, in *Post-Keynesian Economics*, ed. by K. Kurihara, Ch. 8 (Rutgers University Press, New Brunswick, 1954), pp. 192–226
18. R.M. Goodwin, The nonlinear accelerator and the persistence of business cycles. Econometrica **19**(1), 1–17 (1951)
19. A. Krawiec, M. Szydlowski, The Kaldor-Kalecki business cycle model. Ann. Oper. Res. **89**, 89–100 (1999). https://doi.org/10.1023/A:1018948328487
20. A. Krawiec, M. Szydłowski, Economic growth cycles driven by investment delay. Econ. Model. **67**, 175–183 (2017). https://doi.org/10.1016/j.econmod.2016.11.014
21. A. Krawiec, M. Szydlowski, On nonlinear mechanics of business cycle model. Regul. Chaotic Dyn. **6**(1), 101–117 (2001). https://doi.org/10.1070/RD2001v006n01ABEH000168

22. M. Szydlowski, A. Krawiec, J. Tobola, Nonlinear oscillations in business cycle model with time lags. Chaos, Solitons Fractals **12**(3), 505–517 (2001). https://doi.org/10.1016/S0960-0779(99)00207-6
23. M. Szydlowski, A. Krawiec, The stability problem in the Kaldor-Kalecki business cycle model. Chaos, Solitons Fractals **25**(2), 299–305 (2005). https://doi.org/10.1016/j.chaos.2004.11.012
24. Y. Takeuchi, T. Yamamura, Stability analysis of the Kaldor model with time delays: monetary policy and government budget constraint. Nonlinear Anal.: R. World Appl. **5**(2), 277–308 (2004). https://doi.org/10.1016/S1468-1218(03)00039-7
25. X. Wu, L. Wang, Multi-parameter bifurcations of the Kaldor-Kalecki model of business cycles with delay. Nonlinear Anal.: R. World Appl. **11**(2), 869–887 (2010). https://doi.org/10.1016/j.nonrwa.2009.01.023
26. G. Mircea, M. Neamţu, D. Opriţ, The Kaldor-Kalecki stochastic model of business cycle. Nonlinear Anal.: Model. Control **16**(2), 191–205 (2011)
27. L. Zeng, X. Xu, Z. Huang, Complexity in the stochastic Kaldor-Kalecki model of business cycle with noise. Res. Rev.: J. Of Stat. Math. Sci. **1**(2), 1–13 (2015)
28. H. Poincaré, Mémoire sur les courbes définies par les équations différentielles I–VI, in Oeuvre, vol. 1 (Gauthier-Villar, Paris, 1880–1890)
29. C. Baker, C. Paul, D. Willé, Issues in the numerical solution of evolutionary delay differential equations. Adv. Comput. Math. **3**(3), 171–196 (1995). https://doi.org/10.1007/BF03028370
30. R. Bellman, K.L. Cooke, *Differential-Difference Equations* (Academic Press, New York, 1963)
31. H. Smith, *An Introduction to Delay Differential Equations with Applications the Life Sciences* (Springer, New York, 2010)
32. N. Kaldor, A model of the trade cycle. Econ. J. **50**, 78–92 (1940)
33. A.J. Laramie, D. Mair, A.G. Miller, Kalecki's investment theory reconsidered, in *Kalecki's Economics Today*, ed. by Z.L. Sadowski, A. Szeworski (Routledge, London, 2000), pp. 143–161
34. M. Peeters, Investment gestation lags: the difference between time-to-build and delivery lags. Appl. Econ. **28**, 203–208 (1996)
35. U. Brandt-Pollmann, R. Winkler, S. Sager, U. Moslener, J. Schlöder, Numerical solution of optimal control problems with constant control delays. Comput. Econ. **31**(2), 181–206 (2008). https://doi.org/10.1007/s10614-007-9113-3
36. P.J. Zak, Kaleckian lags in general equilibrium. Rev. Polit. Econ. **11**(3), 321–330 (1999)
37. M. Szydłowski, A. Krawiec, The Hopf bifurcation in the Kaldor-Kalecki model, in *Computation in Economics*, ed. by S. Holly, S. Greenblatt. Finance and Engineering, Economic Systems (Elsevier, 2000), pp. 391–396
38. R.A. Dana, P. Malgrange, The dynamics of a discrete version of a growth cycle model, in *Analysing the Structure of Econometric Models*, ed. by J.P. Ancot, part 7 (Martinus Nijhoff, The Hague, 1984), pp. 115–142. https://doi.org/10.1007/978-94-009-6098-5_7
39. M.F. Atiyah, G.W. Moore, A shifted view of fundamental physics. arXiv:1009.3176
40. J. Lei, M.C. Mackey, Deterministic Brownian motion generated from differential delay equations. Phys. Rev. E **84**, 041105 (2011). https://doi.org/10.1103/PhysRevE.84.041105